Praise for *Ripple Effects*

"A marvelously thorough synopsis of the many daunting issues surrounding lake management. Rulseh utilizes case studies, interviews, and a storytelling format to frame the issues, making the book an easy and pleasurable read. He also doesn't leave the reader wallowing in the problems, instead offering sage advice on the many ways we can turn our love of waters into positive actions to protect them. Highly recommended."—John Bates, author of *Wisconsin's Wild Lakes: A Guide to the Last Undeveloped Natural Lakes*

"Illuminates the tight connection between what we do on the land, the health of our lakes, and the health of our economy and communities. This book will forever change how you think about lakes and the Northwoods. And it offers insights into what you can do to protect our beloved lakes, and make them more resilient in the face of environmental change in our rapidly changing world."—Jake Vander Zanden, director of the Center for Limnology, University of Wisconsin–Madison

"In *Ripple Effects*, Rulseh describes in engaging detail the pressures we're putting on our beloved lakes and offers solutions that are within reach of everyday people. He expertly weaves personal experiences and extensive research into an invaluable lake stewardship guidebook for anyone who loves lakes."—Jo Latimore, senior outreach specialist, Department of Fisheries and Wildlife, Michigan State University

"Wonderfully researched and richly detailed. At a time when more and more people are searching out our waters for their livelihood, for recreation and relaxation, and for building cherished memories with friends and family, Ted Rulseh's rich and readable book tells us not only how intimately connected we are to our waters but how the fate of water is our fate as well. In the end, Rulseh's story is a hopeful invitation, providing those who love lakes and are in a position to protect them the understanding and tools they need to reverse these disturbing trends."—Jeff Forester, executive director, Minnesota Lakes and Rivers Advocates

RIPPLE EFFECTS

How We're
Loving Our Lakes to Death

TED J. RULSEH

THE UNIVERSITY OF WISCONSIN PRESS

Publication of this book has been made possible, in part,
through support from the Anonymous Fund of the College of Letters and
Science at the University of Wisconsin–Madison.

The University of Wisconsin Press
728 State Street, Suite 443
Madison, Wisconsin 53706
uwpress.wisc.edu

Gray's Inn House, 127 Clerkenwell Road
London ECIR 5DB, United Kingdom
eurospanbookstore.com

Printed in the United States of America
This book may be available in a digital edition.

Library of Congress Cataloging-in-Publication Data

Names: Rulseh, Ted, author.
Title: Ripple effects : how we're loving our lakes to death / Ted J. Rulseh.
Description: Madison, Wisconsin : The University of
Wisconsin Press, [2022] | Includes index.
Identifiers: LCCN 2021061604 | ISBN 9780299339609 (cloth)
Subjects: LCSH: Lakes—Wisconsin. | Lakes—Minnesota. | Lakes—Michigan. |
Lake conservation—Wisconsin. | Lake conservation—Minnesota. |
Lake conservation—Michigan.
Classification: LCC QH98 .R794 2022 | DDC 577.6309775—dc23/eng/20220413
LC record available at https://lccn.loc.gov/2021061604

I dedicate this book to my late father,
Roger D. Rulseh, without whose influence I might
never have found fascination with lakes, and to my wife,
Noelle, with whom I share a love of the Northwoods lake country.
This book is also for our daughter and son-in-law, Sonya and Chad
Kulow, and their sons, Tucker and Perrin; and for our son, Todd.
May the lakes be here for them, in as good or better condition
than today, long after I have moved on from this life.

Contents

Preface

Nothing I've read better describes the primal attraction humans have to water than these words from Herman Melville's *Moby-Dick.*

> Say you are in the country; in some high land of lakes. Take almost any path you please, and ten to one it carries you down in a dale, and leaves you there by a pool in the stream. There is magic in it. Let the most absent-minded of men be plunged in his deepest reveries—stand that man on his legs, set his feet a-going, and he will infallibly lead you to water, if water there be in all that region.

For some of us, that water is the ocean or the Great Lakes; for some, wide rivers or rushing trout streams; and for others, like me, the clusters of blue-water jewels in the parts of North America sculpted by the glaciers. The waters differ, but the allure is the same.

To say there's magic in it is no hyperbole. I've been addicted to lakes since I was eight, when I spent a weekend with my father and older brother on Duck Lake, in the big woods of Michigan's Upper Peninsula. The straight, rough trunks of two majestic pines framed the sweeping lake vista from the screen porch of the cottage. Down a short path from the porch stairs, waves of tannin-stained water lapped against a beach of coarse sand sprinkled with pebbles of milky quartz. At the water's edge, irregular mounds of brown-on-white foam quivered in the wind.

Until that weekend, I had never seen a bald eagle, stood beneath such tall trees, or heard the distant wail of a loon. The lakes near our home, amid a landscape dominated by dairy farms, were small, weedy pools, surrounded by houses and cottages, many with mowed lawns, barren of trees. At Duck Lake, white pines and spruces, maples and birches towered over the shorelines. Except for a modest log house on an island across a quarter mile of water, I could see no other structures from a vantage point on the beach.

For years after that first visit, my family—Mom and Dad, four boys and four girls—took a week (sometimes two) of summer vacation at the cottage, owned by one of Dad's coworkers. Now, six decades later, I still visit Duck Lake at least once a year, fishing from one end to the other, but always ending at sunset anchored just out from the site of that old cottage (recently demolished), above the weedy rock-and-sand bar where, as kids, we sat in rowboats and soaked bits of nightcrawler for perch, bluegills, sunfish, and bass. I've come to know the lake well, all eleven hundred acres, yet I still discover something new each time I go.

One type of discovery concerns me: every year it seems someone has built a large new house on the shore, in some cases sacrificing the trees for a landscape worthy of a country club golf course. Duck Lake is still far enough north, far enough from the tourist-magnet towns, to have remained suitably wild. Many other lakes are less fortunate. The traditional Northwoods allure largely persists, but each year more wooded lots are bought and built upon; each year more small cabins, in the same families for decades, are sold off and larger houses erected. Along with this trend come added stresses on the lake ecosystems, some quickly obvious, others less so: Nutrient pollution and algae blooms. Invasive species. Failing septic systems. Bigger and higher-powered boats creating wakes that can erode shorelines. A changing climate. Meanwhile, state and local government programs aimed at lake protection are, in many cases, severely understaffed and underfunded.

All this is sobering, and so is the reality that, as a lake property owner and resident, I am part of the problem. I had wanted a place on a Northwoods lake ever since those long-ago Duck Lake summers. My wife and I finally got one in 2009. We had always thought of buying an existing cottage in order to avoid contributing to the rampant buildup of lakefronts. As it turned out, we bought a vacant lot on a wooded slope above Birch Lake in northern Wisconsin's lake-rich Oneida County, and there we built a modest two-bedroom home.

Our sensibilities being what they are, and based on what we had learned over the years from people we considered good lake stewards, we chose to tread lightly, mainly by keeping the woods of white pine, hemlock, balsam, and oak between the house and the lake fully intact. But we're not inclined to nominate ourselves for some kind of stewardship sainthood; we have impacts on the lake by the mere fact of living where we do. In the realm of lake advocacy, there is no room to be self-satisfied.

Meanwhile, amid all the threats to our lakes, there are causes for optimism. Many lake residents have left their properties reasonably natural; long sweeps of shoreline remain wooded, especially on lakes partly within state, national, or county forests. Loons and eagles and other wildlife abound. Fishing is generally good. Water quality in most lakes is still high. There remains a vast treasure of natural bounty and scenic beauty to protect. Most important, armies of lake advocates are at work daily—in state natural resources agencies, in universities and extensions, in county and local governments, and in myriad groups such as watershed councils and lake associations—promoting sound property development practices and good lake stewardship.

Still, for someone in love since childhood with a wilder Northwoods, the trends are concerning. No one is doing damage on purpose; no malignant force is at work to degrade the lakes. No sinister industry dumps in pollutants—threats like acid rain and toxic mercury deposition from electric power plant emissions were largely cured years ago by federal and state regulation.

No, the simple truth is that we who live on the shorelines are loving our lakes to death. Like Melville's Ishmael, we're enchanted by water. We want our lakefront homes, our boats and boathouses, gazebos and lawns, garages and storage buildings. And if we're not careful, we risk setting the lakes on a course for slow, steady, and possibly at some point irreversible degradation.

The fundamental reason we harm the lakes is not that we don't care. It's that we often fail to connect what we do on the land with what happens as a consequence to the lake—to the water's clarity; to the fishes' reproduction, growth, and survival; to the overall health of the lake ecosystem; to the desirability and value of the homes on the lakeshore; and to the quality of lake country life itself.

This book aims to help establish the connection. I've interviewed lake experts of many stripes from around the region and reviewed numerous research reports, all to understand the nature and severity of threats to our lakes, their underlying causes, and the cures available. The information can serve as a guide for how, together, we can enjoy our lakes while providing the optimum measure of protection.

There are remedies for most of what ails our lakes; many simply require changing the ways we live on and around them. Government regulation and financial support surely must play a role, but the remedies in large part depend on all of us doing the right things—one property, one home, one boat, one lake at a time.

Acknowledgments

It is nearly impossible to thank properly all the people who helped make this book a reality. They include the many sources who gave of their time and knowledge for interviews, as well as those who did the scientific and other work that is quoted in the text and cited in the notes. Rather than list those people here, I ask that you remember my debt to them as you encounter their names on these pages.

I am also grateful to many thousands of people I have never met and who are the inspiration for this book. They are lake association and lake district members, state and county natural resource staffers, university professors and extension specialists, and others who work tirelessly, for pay or as volunteers, to protect and improve our lakes. Chances are the lakes you love most deeply are better for the efforts of heroes like these who persist despite formidable obstacles, heeding those immortal words of Sir Winston Churchill: "Never, never, never quit."

It takes special people to know what must be done, meet formidable barriers along the way, and press on regardless. They face chronic shortages of money and personnel. Their vital messages and notes of inspiration often fall on deaf ears. They struggle with the seemingly irresistible forces of lakeshore development and the often immovable objects of state legislatures. They see progress move as slowly and near imperceptibly as the hour hand on a clock. Still they rise day by day, year by year, to pursue their mission.

I am privileged to call a number of such people my acquaintances and friends. They include the members of Crew 11 in the Wisconsin Lake Leaders Institute; the board members of the Oneida County Lakes and Rivers Association; staff members of Wisconsin Lakes and the University of Wisconsin

Extension Lakes team; fellow residents of Birch Lake, where I live; and others I have encountered through volunteer projects on behalf of lakes. The task of lake advocacy is too large and difficult to even consider undertaking except with an army of like-minded people.

Finally, I need to recognize five whose names you will not see elsewhere in this book but who had a great deal to do with the final shape of it. They are the members of my writers group who over nearly two years critiqued the work, chapter by chapter. Seeing it from the viewpoint of my intended audience, they suggested many improvements, large and small. I have learned from experience that to ignore their comments is a big mistake. So, Sue Drum, David Foster, Andrée Graveley, Cheryl Hanson, and Elaine Hohensee, thank you.

RIPPLE EFFECTS

CHAPTER 1

A Lake, a Family

"It's the greatest place on earth."

That's how my cousin Tom Rulseh describes his family's cabin on a small, spring-fed lake in northern Wisconsin's Vilas County. "There's the love and affection that go with a family place provided by our parents," he says. "Everything here is pure and clean, peaceful and natural. How could it get any better?"

His sentiment toward Pine Hill, on forty-one-acre McDonald Lake, will ring true to the many families who decades ago carved out their own tracts of paradise in the northern reaches of Wisconsin, Minnesota, Michigan, and elsewhere in the glaciated northeast quadrant of the United States. One difference is that McDonald Lake has been spared the rampant building, shoreline domestication, motorized sports, and other pressures that now afflict numerous waters. The lake remains much the same as when my uncle Roy first saw it sixty-six years ago. In fact, it's probably better for the environmental ethic and regulations that sprang from the inaugural Earth Day, and for the persistent good stewardship of my cousin's family and their neighbors.

Today, Pine Hill includes nine hundred feet of lake frontage, nine acres of woods, and a two-bedroom cottage that finally, just two years ago, became a fully equipped, four-season cottage. The family built it up, piece by piece, starting with an eight-by-eight-foot shed that Roy nailed together more than five decades ago. Ownership has passed to Tom and his sisters, Elena and Joan. Their children, now adults, visit often and in recent years introduced the place to the babes-in-arms of a fourth generation.

Two years ago on a snowy day in mid-November, I sat at Pine Hill's kitchen table as Tom recounted, over hot cider, the property's history. Roy

The cabin at Pine Hill. (Ted Rulseh)

and his wife, Phyllis, who lived in Milwaukee, learned about McDonald Lake from an in-law. In 1954 they booked a summer week on the lake at River Forest Resort, just a collection of small cabins and a strip of sand beach with a few boats to rent. Tom was four years old then; his sister Elena six.

The family returned to the resort each summer for eleven years, usually for a week, sometimes for two. "On arriving, the first thing we would do is head down to the beach with inner tubes and just get in the water and play," Tom recalls. "The resort atmosphere was always fun. There were other kids around. There was some playground equipment we could use." By day, chipmunks scampered around the grounds; the kids treated them like pets, tossing them peanuts and other treats. At night, raccoons raided the trash cans. Now and then a deer happened by.

In those times, though, Northwoods life wasn't all idyllic. One summer the resort owners, looking to suppress mosquitoes, fogged the area with DDT. "That put the kibosh on a lot of birds," says Tom. "It just got quiet, and I felt bad. I thought, 'That's really not a good thing at all.'" One evening each week the family would hop in the car and take their garbage to the dump, a low spot in the woods a few miles away. They would toss the trash onto the heap and stay to watch the bears come in for a feast. Such dumps of course caused serious pollution and were regulated out of existence. "That's another thing I'm glad we don't do anymore," says Tom.

Tom's favorite times were evenings, after dinner, when he and his dad got into a rowboat and went fishing. "We'd work the shoreline for bass, and my dad would troll a sucker or a big chub out in the deep water and wait for a northern. Invariably we'd get our limit of bass and one or two northerns, some of which could lie end to end on this table without any trouble."

Their preferred bass bait was mud minnows, purchased at a shop in nearby Eagle River. "We could try other baits and not get much activity, but the mud minnows were really attractive to the smallmouth and largemouth bass. We'd put a minnow on a number 4 hook with six- or eight-pound monofilament line, cast it right up to the edge of the lily pads, and let it drop. Sometimes there was a hard hit. Sometimes it was just gradual; you could see the line moving. You'd let them take it just a little bit and then set the hook. It was fun stuff."

The family, including daughter Joan, born in 1957, fell quickly and deeply in love with the lake. Tom still treasures one poignant memory. "My dad was rowing us out to the fishing spot across the lake where we always liked to start. It was a beautiful summer evening, the sun just setting, and as he was rowing my dad said, 'Just think. This is here every day.' I got to thinking about that more and more, and he did too: We could be there a lot more often."

As fate would have it, the resort owners decided to close shop, sell the cabins, and divide and sell the adjoining wooded properties. In 1965 Roy and Phyllis bought two lots—six acres and six hundred feet of frontage—on the lake's northeast corner for $2,500.

"The first summer we put in a road," Tom recalls. "We cut it in by hand, and then a bulldozer came through and smoothed it out a little bit. Originally we tented on the property. Our parents had remodeled the kitchen at our home in Milwaukee, and they kept some of the old metal cabinets. They brought them up here, and we hung them on trees. Then we set up a Coleman cookstove and a couple of two-by-sixes to make a little bench area. That was our kitchen. That's how we enjoyed the place those first summers."

In the early years they had a seven-horsepower Johnson outboard to mount on the family rowboat. "We kept it here and we used it," says Tom, "but in retrospect it was kind of silly. You'd start it up and go a little bit, and then you had to shut it off because you had arrived."

Roy built the original shed as a place to store everything between visits. Then one day on arrival, the family had to pull out and set up the tent in a heavy rain. So Roy doubled the shed to eight-by-sixteen feet to provide shelter and, in dire conditions, a place to sleep. He put in foam-rubber mattresses, moved the kitchen cabinets inside, and added a small sheet-metal wood-burning stove to take the chill out when needed. Next came an electric line from a utility pole, a septic system, and a well that Roy and Tom spent an afternoon installing by hand.

Roy, an electrical and combustion engineer with precollege experience as a pipefitter's assistant in the shipyard at Manitowoc, Wisconsin, was fully comfortable with the well project. "We went down the hill to about thirty feet from the edge of the lake," recalls Tom, then age sixteen. "We took a section of two-inch galvanized pipe, put a sand point on the end, and just started driving it down with a sledgehammer. We whaled away at that thing, taking turns."

At first each hammer stroke drove the pipe down about an inch; after about ten feet the progress slowed to half an inch per stroke. "Luckily, the glacier didn't leave any rocks in that vertical column," says Tom. They added section after section of pipe and kept hammering the point down through the sandy earth, stopping now and then to rest and check for water with an eight-penny nail hung on a string. Drop it down, bring it up . . . dry. *Clank. Clank. Clank.* Drop it down, bring it up . . . still dry.

Until finally, "Aha, it's wet!" says Tom. "We drove the pipe a little bit deeper. Then we put on a hand pump, primed it, and *whoosh*, up came this cold, crystal clear water. It was just beautiful." The finished well ran twenty-four feet deep. For a few years family members walked down the hill to fetch water in buckets. Then Roy installed a pressure tank and electric pump to deliver water up to the cabin.

Meanwhile, the cabin expansion continued. Another eight-foot-square extension made it L-shaped, sixteen feet on the long sides. Next came a living room, then a bedroom, which Roy's younger brother Roger helped him build. More furniture. Kitchen appliances. A more serviceable cast iron wood stove. Bit by bit, life at Pine Hill became more comfortable.

In the late 1980s, a parcel adjoining the property went up for sale, and Roy and Phyllis bought it for $9,000, adding three acres and three hundred feet of lake frontage. After a few decades with the same employer, Roy had accrued several weeks' vacation, much of which he and Phyllis spent at the cabin. "Dad jokingly called it his place of work," says Tom. "When we were at the resort, it was just play and fish. At Pine Hill, there was always another project to do. But Dad enjoyed it all."

Phyllis loved it too. She had grown up in the country and liked being outdoors; it was she who named the place. "She really enjoyed her time here," says Tom. "She never got into the building projects, but she was always there with a glass of water or lemonade to encourage the builders, to keep us hydrated so we could keep going."

Tom became one of the builders as he came of age. At seventeen he started driving to Pine Hill on his own, usually when his parents were there, to help his father with projects. One of those was building a stairway down the hill to the lake. Tom was an Explorer Scout then and his father, the Scoutmaster, invited some other Scouts for a visit. "It was kind of a devious plan," Tom recalls with a grin. "Here's a project, guys." With shovels, wooden boards, and metal stakes, the willing conscripts fashioned twenty-three earthen steps that remain to this day.

Roy's work went well beyond steadily improving the property. In 1972 he organized the McDonald Lake Association, consisting mainly of people who owned cabins that had been part of the resort. "It sprang from an awareness that we were sharing this body of water, and we all had different interests," says Tom. "We talked about those interests to make sure we

McDonald Lake, seen from Pine Hill. (Ted Rulseh)

were doing things we all found agreeable, and doing them with the lake as the focal point, because that's why we were all there."

The group's first consequential decision was to restrict waterskiing. "We asked our members: What does the lake mean to you? For many it was fishing. For some it was swimming. For some it was waterskiing, but they were definitely the minority. It's not a good waterskiing lake—you'd just go around and around in tight little circles. First we thought maybe we could limit the hours, so from 11 a.m. to 3 p.m. would be waterskiing time. Eventually we agreed we'd rather not have waterskiing at all." Soon afterward the state Department of Natural Resources declared lakes smaller than fifty acres to be no-wake only. Quiet still reigns on McDonald Lake.

A few years later the association acted when an eighty-acre property next to Pine Hill went up for sale. There was talk of the owner turning it into a mobile home park with keyhole development: one pier or marina to which multiple lot owners would have access. Concerned about how that would affect the ambience and ecological character of the lake, the group successfully petitioned the Town of Cloverland, and ultimately Vilas County, to change the land's zoning to single-family.

Several years ago, the McDonald Lake group joined other lake associations in the township to fund a lake quality analysis performed by a management consultant. The aim was to create a baseline study of aquatic plants and an analysis of the surrounding watershed, as a reference point for monitoring changes in the environment. That study added to data Tom has collected annually since 1986 as a volunteer for the Citizen Lake Monitoring Network, a state-sponsored program in which people measure and report water-quality parameters. The data covers water clarity, temperature profile, chlorophyll, and phosphorus: "It's all part of our effort to say, 'We love this lake. What can we do to keep it healthy?'"

As Pine Hill matured, the extended family grew. Tom married Vicki Reuling, his college sweetheart, and eventually, with their son, Jonathon, they moved to the community of Three Lakes, just over a dozen miles from the cabin. Elena and her husband, Zaza, visited the lake regularly, as did Joan with her husband, John, and their children, Ted, Peter, Maggie, and Anne.

Family life at Pine Hill conjured memories of early years on the lake. "It was a natural place to come with our kids; an extension of how it all started, when we used to come to the resort," says Tom. "Jonathon would be here with his cousins playing in the water. It was a very similar experience, except now we had our own place. After spending much of the day at the waterfront, we'd come up to the cabin and have dinner. At dark, we'd get out the board games, and the kids would gather around the dining room table and play."

To make more lodging space, Roy built what the family calls the Annex, a small cabin with a bedroom, a bathroom, and a workshop. He also added a sixteen-by-eight-foot screen house intended for the grandchildren. "He didn't want a bunch of screaming kids around all the time," says Tom. "He thought they should be outdoors more. But the kids were afraid the bears would get them, and they would never sleep in the screen house. So now Vicki and I sleep there when we visit in the summertime and other families are here."

Along the way, Pine Hill has survived two traumatic events. In summer 1999, an intense thunderstorm spawned a violent downburst that swept across the lake. "In a convection storm, there are big cumulus clouds that contain a lot of turbulence and updrafts," says Tom, calling on his college degree in geography. "An updraft goes only so high, and the subsequent downdraft can reach eighty to a hundred miles per hour. When it hits the ground, the wind travels along the surface."

The wind cut a swath about 250 feet wide through the Pine Hill property, felling hundreds of trees, including thirty-inch-diameter pines, some knocked clean over, others snapped off like twigs halfway up the trunk. All the downed trees lay pointed in the same direction, some blocking the road to the cabin. None of the buildings were seriously damaged. "We hired a tree service to come in and cut the timber," Tom says. "With those big trees lying on top of each other, we thought it was too dangerous to do by ourselves."

Seven years later, on a windy day during a late-spring dry spell, Tom received a phone call alerting him to a report of a fire at Pine Hill. "They said, 'Don't go there,'" Tom recalls. "I said, 'I'm going.'" He and Vicki took an alternate route to the property to find the fire mostly extinguished and a DNR crew and the Eagle River Fire Department still at work, using a bulldozer to clear away trees. "To the fire department's way of looking at it, the fire was out, but a couple of pines were still smoldering, with glowing coals inside the trunks. I started getting buckets of water and throwing it on."

The fire started when a tree blew down in the wind, hit the power line feeding Pine Hill, and caused a spark that lit off dry tinder. Someone saw the smoke from across the lake and called the fire department, likely saving the property. "We lost about three-and-a-half acres of trees," says Tom. "The fire came right up from the bottom of our hill. The pines were all healthy, coming back after the windstorm, and some that weren't affected by the storm were burned. I thought, 'We're going to have wind events, and we're going to have dry weather. We don't want to worry about this anymore.' So we buried the power line to eliminate that hazard."

Pine Hill made it through those trials, and lesser events such as a beaver assault on shoreline trees, and lake water levels falling sharply in times of drought, then recovering during wet years. As always, eagles soar overhead. In most years the lake hosts a nesting pair of loons. Fishing remains productive, the largemouth bass robust and abundant. The cabin has a new foundation, new insulation and siding, and a variety of interior improvements, along with reminders of times past, such as the old dining room table and a 1930s vintage Tru-Cold refrigerator.

Pine Hill is a place where, each year, prized rituals are observed and new chapters written in the life of an extended family. Always at the center, there's the lake, in many ways unchanged since the glaciers retreated and left it sparkling in the sun. Says Tom, "We feel really lucky to be here."

Multigenerational cabin owners elsewhere feel fortunate, too—even as current and future stresses on lakes increasingly threaten their legacies and dreams.

CHAPTER 2

Cabin Country

On a November Saturday night, snowflakes in the air, the final seconds tick down at TCF Bank Stadium in Minneapolis. The Wisconsin Badgers football team has beaten the Minnesota Golden Gophers, 38–17. When the scoreboard clock hits 0:00, the Badgers in their red-on-white uniforms spring from the sideline and sprint toward a goalpost where leans the game trophy, Paul Bunyan's Axe. As the team gathers around, players take turns grabbing the oversized axe and pretending to chop the goalpost down. It's a reminder of the Big Woods logger who, legend has it, cleared the northern pinery with help from his lumber camp crew and his partner, Babe the Blue Ox.

For decades, Wisconsin and Minnesota have laid energetic claim to Paul Bunyan, and in fact so does a neighbor to the east: the annual Michigan/ Michigan State football game is played for the Paul Bunyan Trophy. Across these states and others, the Bunyan legend lives on in statues, restaurants, a campground, a state trail, a museum, a resort, and more. The annual football games of course settle nothing.

Paul Bunyan and his lumberjacks didn't clear the forests. Real loggers did, and they left behind, amid the stump-strewn land, the thousands of lakes (said to be Babe's hoofprints) that now are havens for anglers, hunters, skiers, paddlers, and all manner of tourists taking breaks from the rigors of work and city life. Over the years, the states' serious competition has been not for the Bunyan legend but for the attention and the bankrolls of those visitors.

The forests in the glaciated northeast quadrant of the United States surround some of the most extensive clusters of lakes in the world. I live on one of those lakes—180-acre Birch Lake in north central Wisconsin's Oneida County. That lake's shoreline and surroundings developed much the same way the entire North Country grew up, starting with small seasonal cottages, campgrounds, parks, and rustic resorts. My neighborhood's history vividly illustrates how an entire lake-rich region became a place that rightly could be called cabin country.

As for the logging era, the stories across the lake country of Minnesota, Wisconsin, and Michigan are much the same. Explorers and settlers coming to the region in the 1830s usually did not encounter pure stands of pine. Huge white pines were the most commercially valuable, but they mainly grew just a few dozen to the acre, towering over mixed woods composed of hemlock, sugar maple, and yellow birch. Only about 10 percent of Wisconsin's presettlement forest was pine.

From the 1830s until the early twentieth century, the loggers cut the white pine for lumber and sent the logs down the rivers to the mills. They also harvested the hemlocks for their bark, from which tannic acid was extracted for the leather tanning industry. Later, the hardwoods were logged

The cutover often left behind a barren landscape. (courtesy of Forest History Society, Durham, NC)

off and shipped to market, mainly by railroads built to serve the timber and mining businesses. Finally, the less economically valuable trees were cut. By the 1930s, most of the marketable timber had been removed.[1]

Farming Futility

The decades of logging left behind a wasteland. Botany professor John T. Curtis writes: "In many places the entire landscape as far as the eye could see supported not a single tree more than a few inches in diameter. Only the gaunt stumps of the former pines ... remained to indicate that the area was once a forest rather than a perpetual barren."[2] That desolation led to catastrophic fires. Logging left behind volumes of slash—branches and fallen discarded trees—that dried in the sun and were lit on fire by causes from lightning to sparks from passing steam locomotives to fires set by farmers clearing stumps from logged land. Many fires covered large areas, leaving behind ashes, consuming organic matter from what had been the forest floor, and damaging the topsoil.

During the logging era and even for some time after, little thought was given to conservation. For the first three decades of the twentieth century, the consensus was that the plow would follow the axe. State governments, newspapers, merchants, bankers, real estate agents, and academic experts promoted the cutover lands for conversion into small farms, leading to an intense but short-lived land boom. Logging companies sold tracts to speculators who in turn sold individual parcels, many of them to immigrants from Europe.[3]

Early in the new century, thousands of settlers moved into the region to farm the cutover land, drawn by relatively low prices and favorable credit. But most of those farms were doomed from the outset. It was arduous work to dig out or dynamite the stumps and clear away the native rock before cultivation could begin. Accordingly, many farmers worked around the obstacles, engaged in limited dairy farming and potato growing while making ends meet by hunting, fishing, and taking jobs off the farm. Furthermore, the growing season was short, and most of the region's soils were too infertile for crops, some of it so sandy as to make profitable farming impossible. Naturalist John Bates writes, "Add in the North Country's very inhospitable winters and the great difficulty isolated homesteaders had to transport products to urban markets, and the democratic dream of a nation of small farmers dissolved in the ecological and economic realities of the Northwoods."[4]

As the farms failed and the settlers moved on, millions of acres of land were left deserted and tax-delinquent, placing huge burdens on towns and counties. Bates writes, "And then came the coup de grâce in 1929—the Depression. Some northern towns reported tax delinquency at 70 percent of all their land, while many whole counties reported up to 50 percent of their land as tax-delinquent. The plow had not followed the axe."[5]

Forests Reborn

With the end of the farming era, the woods in time regenerated; the region is again home to healthy forests and a variety of wooded habitats, although the species mix has changed since before the land was logged. The transition was neither fast nor smooth; it required a combination of natural processes and extensive government intervention. The cutover had radically transformed the landscape. Without tree and plant roots as anchors, much topsoil was lost to erosion. Trees like aspen, paper birch, and jack pine that thrive in bright sunlight and poor soil sprouted by the millions, providing habitat for animals that prosper in the early stages of woodland succession. Chief among these were white-tailed deer, which fed on white pine seedlings and so inhibited regrowth.[6]

As early as the first decade of the twentieth century, movements were launched to promote forestry and to establish forest reserves, primarily as sources of timber for harvest—a different kind of agriculture. Those initiatives met resistance as farming in the cutover regions continued, and as county and town governments rebelled against the prospect of placing vast stretches of land under government control, removing it from the property tax base. But in time the reality took hold that the land was best suited for producing trees. Bates writes, "It took a total failure of the farm dream before enough hard-earned despair forced a change in the cutover mythology. Most people and politicians now had no choice but to favor county, state and federal ownership of the forests. . . . Culture eventually bowed to nature, as it always must, and the burned-over lands were growing trees again."[7]

The early decades of the twentieth century saw the beginnings of national, state, and county forests. The United States Forest Service was organized in 1905, and the national forests were created from 1909 to 1938. In the 1930s and early 1940s, workers in the Civilian Conservation Corps, a New Deal program, planted hundreds of millions of trees while fighting fires, building

Forests regrew naturally after the cutover, and tourism followed. (courtesy of
Forest History Society, Durham, NC)

roads, improving streams, stocking lakes with fish, and constructing park
buildings.[8]

Reforestation took hold across the region, and the forests gradually recovered enough to once again enable profitable timber harvesting. By now,
the need for forest conservation and sustainable management practices
had become established. Government agencies and foresters from private-
sector companies began collaborating to help the forests grow.

Natural Treasures

The forest and lake country naturally began attracting tourists, and state
governments and regional business associations aggressively promoted its
charms. Railroads originally built to serve logging and mining now brought
in tourists from cities to the south, some trips promoted as "fisherman's
specials." In the 1920s, new roads and improving automobiles made the
lake country more accessible. For many residents of Minnesota, Wisconsin,
and Michigan, "the joys of fishing, hunting, hiking, camping, and other

forms of outdoor recreation were as close as an afternoon drive. The popularity of these activities increased throughout the decade as members of
all socioeconomic classes began to participate."⁹

By the 1930s, the states were doing vigorous battle for their shares of
the tourist trade. State slogans were born. Minnesota touted "The Land
of Ten Thousand Lakes." The Badger State invited visitors to "Relax in Wisconsin, Where Friends and Nature Meet." Michigan offered a "haven of
contentment" where "climate, scenery, lakes, streams, abundant wildlife
and an excellent system of highways satisfy any taste."¹⁰ State parks were
created and roadside parks and waysides were built to give motorists pleasant respites on the way north. State governments and groups like the Minnesota Arrowhead Association and the Upper Peninsula Development Board
promoted tourism through brochures, newspaper and magazine ads, dis-
plays at sport and travel shows, and offices in cities like Milwaukee, Chicago, Cleveland, Minneapolis, and Detroit.

Before long the region was home to numerous campgrounds, hundreds
of small lakeside resorts with clusters of housekeeping cottages, and thousands of one- and two-bedroom seasonal cabins. Tourism in the lake country was coming of age. It was an opportune time for an insurance agent
named Art Thompson, from southern Wisconsin, to visit lake-rich Oneida
County to the north and discover Birch Lake and Sand Lake, separated by
a strip of land topped by a glacial ridge. It was he who surveyed the land
to create the lakefront lots and the neighborhood my wife and I now call
home. We purchased our lot from Jim Thompson (one of Art's grandsons)
and his wife, Mary. Another grandson, our across-the-road neighbor Denny
Thompson (Jim's older brother), described the area's transformation from
reforesting cutover land to a loosely knit, lake-centered community.

Crystal clear, spring-fed Sand Lake (37 acres) fed a creek that flowed into
Birch Lake (180 acres). Birch Lake was called Artie Lake while the loggers
were active. It later became Shepard Lake and then Ruth Shepard Lake; it
acquired its current name in 1920.

Art Thompson learned of this place in the early 1930s from a friend who
drove a taxicab in Milwaukee but spent his summers operating a tavern and
small resort nearby. Denny recalls, "He told my grandpa, 'You could make a
living up there. There are all these lakes. You could hunt and fish.' Grandpa
loved all that stuff." In 1938 the friend quit his taxi business to run his
resort full-time. In that same year, Art Thompson moved north and, after
exploring the area, found a farmer with land for sale. On a land contract, he

bought twenty wooded acres with eight hundred feet of frontage on Sand Lake and twelve hundred feet on Birch Lake, all for $920. By 1941, he had built a house for himself and two resort cabins on Birch Lake's northeast shore, near the mouth of the feeder creek. He gradually built it up to six cabins, each with a metal rowboat, a set of oars, and an anchor fashioned from a coffee can filled with concrete. Meanwhile, tax-delinquent land could be bought for a song, and over the years Art purchased parcels in the area totaling some two thousand acres, for as little as sixty-three cents per acre.

The Resort Life

Art finished building the resort in 1946. Birch Lake Resort, as it was called, wasn't the first on the lake. As early as 1916, taverns and small resorts came and went and, in some cases, burned down. Denny Thompson's extensive historical records show nine resorts operating on Birch Lake at one time or another. The resort experience was bucolic. Most resorts had a small stretch of sand beach. Guests hand-pumped water from sand point wells and used outhouses. Septic systems, if they existed, were little more than fifty-five-gallon barrels with drilled holes, buried in the sandy soil. "For electricity, some resorts had 32-volt Delco-Light or Kohler sets generating their electricity, along with a battery bank," Denny says. "If you had one battery, you could use your lights. If you had six batteries, you could run a resort or a restaurant. Art ended up with five batteries. When they ran down, the generator would start by itself and charge them up."

In the 1940s, ten dollars would buy a week in a one-bedroom house-keeping cabin. "If you had a Murphy bed and another bed, that was fifteen dollars a week," says Denny. "You brought your own food. The farmers would come around and sell you fresh milk and eggs. They'd go door to door. All the little resorts had rowboats to rent. It was quiet. The first motor-boat and the first water skier on Birch Lake came in 1927."

Running a resort was a hand-to-mouth existence, and owners took all steps necessary to earn extra money and hold down expenses. "Before the assessors came around once a year," says Denny, "Art would pull his boats over the hill, put them in the brush on the other side of Seed Lake, and hide them. Otherwise he would have to pay a personal property tax of five dollars per boat, which in those days was a lot of money." Art supple-mented the resort income by keeping some of his insurance business, train-ing and selling hunting dogs, selling hunting rifles, and buying and selling tax-delinquent real estate.

Denny recalls visiting the resort as a kid with his parents and four siblings. At the time, in the late 1940s, Birch Lake lived up to its name; it
was surrounded by paper birches. "It was all white," Denny recalls. "Some
big pines that had survived the fires were still there. It was really pretty."
The kids would swim at the resort and also in Sand Lake with its smooth,
sandy bottom. They often camped on Birch Lake at the east end of Art's
property, at the bottom of a sandy downslope where mostly level lake frontage made it easy to drag canoes in and out. Once in a while a bear lumbered
by, frightening the kids and putting their dog in a panic.

Birth of a Neighborhood

Art Thompson ran his resort until the early 1950s, then put the cabins
and accompanying ten acres up for sale. He kept his remaining ten acres,
essentially a 400-foot-wide strip of land between the two lakes. In 1956,
the electric utility installed a power line along a trail that ran through the
middle of the property. By that time, land values were rising, and Art plat-
ted and subdivided his property, most of the lots one hundred feet wide,
extending from one lake to the other. He then began selling the lots for
a few hundred dollars each, generally just one or two per year, as a source
of retirement income. Among the first buyers were Denny Thompson's
parents. Denny recalls:

> After that we came up here a lot. We built our first cabin on Sand Lake.
> We dug out the basement by hand. Dad had a store in West Allis [a Mil-
> waukee suburb] and worked there during the week. We'd drive up on Fri-
> day night and get here sometimes at midnight. We'd work on Saturday, go
> to church on Sunday morning, and then go home. All we did was work.
> We'd cut down trees, dig basements, put in wells. If we got here before
> dark on Friday during those long days in summer, we'd go fishing on Birch
> Lake. At night sometimes we caught bullheads on Sand Lake. Then on
> Saturday night we'd have a fish fry.

Denny spent three years in the U.S. Army. Soon after discharge in 1964,
he bought one of his grandfather's lots for $800; eventually his brothers
and a brother-in-law bought parcels of their own. "Grandpa had all of us
paying him twenty-five dollars a month for as long as it took, with no interest," Denny says. Over time, the brothers built their homes, piece by piece,
all on the Sand Lake side. The parents moved into the much-improved and

This cabin on Birch Lake was part of the Art Thompson resort; it has been remodeled and slightly expanded. (Ted Rulseh)

expanded family cabin in the 1980s. Gradually the remaining lots were sold off. The new owners ultimately split the lots that extended from lake to lake, deeding half to a relative or selling half on the market. The Thompsons gradually converted the trail between the two lakes into a graveled easement now named Art Thompson Road. After selling the last of his lots, Art and his wife moved away; he continued to visit until 1973, when he died.

All the while, the rest of Birch Lake and Sand Lake built up in generally similar fashion. Rail service to the area ended in 1956; the tracks were removed in 1980. The Town of Cassian established its first volunteer fire department in 1960. Families from cities a few hours south, many of them former resort visitors, bought lots and built cabins for weekend and week-long summer escapes.

Resort life through the 1960s and 1970s was still idyllic, couples with children getting away for a week or two, anglers connecting for vacations, patronizing the area's bait shops, stores, restaurants, and taverns. The forests around the two lakes matured, here and there red pines and white pines lording it over the canopy, providing perches for bald eagles. Loons arrived with ice-out in spring, calling out for what surely must have been joy. Deer slipped out from the trees to drink from the lakes. Otters inhabited the woods and could be spotted frolicking in the water, or doing their brand of backstroke while crunching on mussels or crayfish plucked from the lake bottoms.

Birch Lake's shoreline eventually was split into lots, most with a hundred feet of frontage; about 10 percent of the shoreline, on the east side, touches the state-owned Northern Highland American Legion Forest and remains undisturbed. The vast majority of roughly a hundred private properties had small, seasonal cabins, some still owned by their original families.

One by one the resorts closed. Soon the only reminder of the resort era was Birch Lake Bar and Resort with its several housekeeping cabins. The birches, short-lived trees, slowly died and toppled to the forest floor, until just a brave few stood stark white against the deep green of pine, hemlock, balsam, maple, and red oak. In short, Birch and Sand Lakes were much like thousands of other lakes all across the former logging territory. But for those lakes, another era of change, and of significant threats, was in the offing.

CHAPTER 3

Paradise Discovered

1980

Little St. Germain Lake Cottage. Cozy 2-bedroom, knotty pine interior, screened porch, nice sandy beach. Gradual slope to lake. $31,900.[1]

We had a dream.

Noelle spent the first two years of her teaching career in the Brainerd Lakes area of Minnesota. I grew up vacationing in a rustic cabin on a wild lake in Michigan's Upper Peninsula. We had Northwoods in our blood. So from courtship into married life, and as our kids grew, we took a week's vacation each summer in a rented lakeside cabin somewhere.

And dreamed.

In time we found a place we liked well enough to return year after year, Jung's Cottages on Birch Lake, in the heart of northern Wisconsin's lake country. Through each grueling year of work, I looked forward to that one blessed week in August, away from the telephone, beyond reach of my boss and clients, deep undercover. I especially treasured the Saturday afternoon arrival at the little two-bedroom cabin called Lakeview.

As soon as I parked the car, Sonya and Todd would grab their duffle bags, rush inside, and emerge in their swimsuits. The screen door, *bang*. Hurried barefoot steps down the railroad-tie stairway, onto the pier, and into the water, *sploosh*. For a while I would watch them from a chair on the oak-shaded deck, each breath of pine-scented air like a nip of elixir. Then I would help Noelle carry in the suitcases, toy and book bags, sewing machine, and other niceties with which we had loaded nearly every cubic inch of the car's interior.

While Noelle organized the cabin, I would launch the fourteen-foot aluminum boat with ten-horsepower Mercury outboard, then go for a swim myself, entering the lake with a mad dash along the pier and the "Cowabunga!" dive that made the kids laugh. After we enjoyed a modest dinner of frozen pizza, I would get out my fishing gear, sit on the deck, and watch the loons while slowly fastening the reels to the rods, feeding the monofilament through the guides, and tying on the lures and bobber rigs on which I had pinned my hopes for the evening.

All week we would stay away from town except for a Sunday morning grocery mission, a breakfast at Paul Bunyan's Cook Shanty, and one obligatory foray into Minocqua's downtown to act like tourists, browsing the shops and carrying home a pound of fudge and a sack of two dozen flavors of taffy from Dan's Minocqua Fudge. Most of the time we just reveled in the cabin and the lake. I would seize on intervals of "opportunity fishing," before the others woke in the morning and for an hour or two up until and past sunset. Then I would come in and read from one of the three or four books I had brought, before and after tucking the kids into bed.

We ate simply: easy-to-fix meals like tacos, deli sandwiches, brats cooked up on the grill outside. Of course we fried up the walleyes, perch, and bluegills I had managed to catch. The kids spent much of their time in the water. Noelle sat at the kitchen table sewing patchwork quilts while looking out on the lake through an expansive multipaned window. I joined the kids in the water, took them fishing or for rides in the dented aluminum canoe that came with the cabin, and sat on the bench at the end of the pier sipping a beer or old-fashioned. In other words, the life idyllic.

After dark, the kids in bed, Noelle and I would sit by the fireplace, where pine logs crackled, and wonder how it would be to have a cabin of our own. And so, sometime during the week, we would put the kids in the boat and take a slow cruise around the lake, close to shore, to find For Sale signs. By then, the late 1980s, it was strictly window-shopping, because prices for lake frontage had spiraled hopelessly out of our reach. As the years went by, the northern lake country was changing around us, inexorably and in disturbing ways.

1985

Upper Buckatabon Cottage. 112 ft. of good sand beach. No bank, stately
pine trees. 2 bedrooms, land contract. $39,900.

I had started shopping for a lake cabin in about 1985, ten years after college.
My family no longer took those Upper Michigan vacations, but I still traveled to the northern lakes each June on a four-day fishing weekend with
hometown pals. On a trip through town, I would stop and grab one of the
real estate booklets and, back at our campsite or rented cottage, leaf through
it. In those days, an old two-bedroom cabin on a lake was affordable even to
people of fairly modest means. My means then, on my weekly newspaper
editor's salary, were less than modest, and so I hatched a plan: What if my
parents, my seven siblings, and I all chipped in and bought a place like that
to share? I asked them. None showed interest, and that was the end of it. I
couldn't buy a place myself, and even if I could have, my vacation time was
too limited to enable proper use of it, unless I wanted to drive up and back
on weekends, five hours one way—not for me. So I tucked the idea away.

And then, just a few years later, it happened. It turned out thousands of
people had dreams like mine, people much further advanced in their careers,
with four, five, or six weeks of vacation, and with a great deal more money
than I had. Cabin prices responded. Sandra Ebben watched it happen as
a real estate agent based in Rhinelander, the seat of Wisconsin's Oneida
County, with more than eleven hundred lakes. "When I first got my license
in the late 1970s," she says, "a typical two-bedroom Northwoods cabin on
a lake brought probably thirty thousand dollars or forty thousand dollars,
maybe fifty thousand dollars. But gradually the prices started to go up." The
major escalation began in the late 1980s and largely continued until the
Great Recession in 2008–9. To no surprise, prices for undeveloped lake
frontage followed a similar pattern, as shown in table 1 (the lower prices in
2009 and later largely reflect the recession's impact).

Various factors drove the demand. "As people got more disposable income,
they became more interested in buying something up north," says Ebben,
now manager of the Rhinelander office of First Weber Realtors. "Then the
infrastructure changed, and it became easier to get up here." Heading north
to Rhinelander from Milwaukee or Chicago once meant taking a state highway that wound through the Menominee Indian Reservation and through
every village and city on the way, each with its traffic lights, stop signs, and
twenty-five-mile-per-hour speed limits. The expansions of State Highway

Table 1. Northern Wisconsin vacant lake lot sales, 1999–2009

Year	Number of sales	Average sale price	Average front footage	Cost per foot of frontage
1999	173	$72,847	192	$379
2001	151	$89,375	201	$445
2003	186	$109,016	181	$602
2005	172	$146,789	192	$745
2007	132	$173,947	188	$860
2009*	77	$152,188	194	$784
2012	79	$125.127	174	$719
2015	92	$125,667	170	$739
2018	130	$109,592	183	$598
2019	113	$115,531	180	$641

*First year after Great Recession
Source: Information from Greater Northwoods (Wisconsin) Multiple Listing Service, representing seven lake counties in the lake country. Data compiled by Kyle Zastrow, appraisal specialist based in Rhinelander, Wisconsin. Includes lots with 75–500 feet of lake frontage, lot size 20 acres or less. Excludes unnamed lakes.

29 and U.S. Highway 51 to four-lane expressways opened an alternate and more westerly route, bypassing the towns. "Now you could get up here a whole lot faster," says Ebben. "That drew more people, and they had more money to spend. It started to snowball. Once paradise got discovered—and that's really what happened—things started to change."

1990

South Two Lake Cottage. Charming cottage with knotty pine interior. 2-bedroom, newer exposed basement located on a crystal clear lake. $45,000.

It was no different in the neighboring state to the west: "You know it's summer in Minnesota when it's early on a Friday afternoon and northbound traffic on I-35 is backed up out of the Twin Cities and both ways on I-94," begins an article in the MinnPost online publication. "Inside the cars are families, fishing poles, coolers, life jackets and other implements for a weekend at the lake."[2]

Prices didn't rise uniformly. For example, a property on a large, clearwater lake cost more than a similar property on a small lake with brown, tannin-stained water. Places on some more prestigious lakes, like Trout Lake and the Minocqua Chain in Wisconsin, Mille Lacs Lake and Gull Lake in Minnesota, and Torch Lake and Lake Charlevoix in Michigan, commanded higher prices. Furthermore, according to Kyle Zastrow, a real estate appraisal specialist in northern Wisconsin, properties on chains of lakes were generally worth more than those on individual lakes. Regardless, buyers bid up the prices of lake cottages.

As prices rose, family lakeside resorts, northern tourist staples with small housekeeping cabins, became casualties. Resort owners found themselves living a difficult, labor-intensive, and not very lucrative lifestyle, with escalating property taxes, while suddenly sitting on long stretches of lake frontage worth a million dollars or more. Meanwhile, vacationers' expectations and lifestyles changed. They wanted better accommodations. Their children were growing and getting more involved in summer school, sports, and other activities. So families were looking not for week-long vacations but for four-day weekends or three-day midweek stays. That didn't fit with the resorts' traditional Saturday to Saturday rentals. "All of a sudden, the economics of those resorts weren't what they used to be," Zastrow says. "How do they make up for that income? Maybe they had to sell off a unit every other year." And sell they did.

Some sold their entire properties to developers, who carved them up for lake lots with one hundred or two hundred feet of frontage and tore the cabins down. Others sold individual cabins as condominiums—often a win-win for the owners and for people with dreams of a lakeside getaway. "As property values went up, the buy-in price point to get a Northwoods cottage kept increasing," notes Zastrow. "Condo conversions were a way to make property on very nice lakes available to a broader segment of the public at more affordable prices, albeit in some cases with marginal or average cottages." And for the owners, selling cabins didn't mean forgoing rental income. They could put those units into a rental pool and offer them to tourists for the weeks when the owners were not there, splitting the proceeds with the owners. In fact, says Zastrow, such arrangements began as early as the 1980s and represented a precursor to the online vacation rental movement.

1995

Squirrel Lake. Yesteryear log cottage, 2 bedrooms, level lot, new drilled well, brand new foundation and some lovely landscaping. Fully furnished with boat and motor. $84,500.

Minnesota's experience illustrates the decline of the family resort. The Minnesota Department of Revenue reported that the number of resorts in the state declined by almost half, from about 1,400 in 1985 to 760 in 2015.[3] That decline continued: The number of Minnesota resorts declined by 3.6 percent in 2016, by 3.2 percent in 2017, by 1.4 percent in 2018, and by 3.3 percent in 2019, leaving just 683 resorts.[4] A major share of these resorts lost were small mom-and-pop enterprises.

The financial advantage to the resort owners is clear. Nick Leonard, a resort owner and an administrator for Otter Tail County, observed that when resort owners consider their options, "and they ask themselves, do I sell this as a resort and walk away with $750,000? Or do I parcel it out and sell it as a common-interest community, essentially an association, and I walk away with $1.5 million, it's a no-brainer what option they're going to pick."[5] The same easy logic applies in another scenario: "if I can sell my resort for $3 million to a developer . . . or $1.5 million to a young couple that wants to start in the resort business."[6] For many resort owners, the sale of the property represented a reward in retirement for a lifetime of work and dedication.

Meanwhile, escalating prices gave many individual owners of cabins or vacant, wooded lake properties a powerful incentive to sell. The rising prices did not deter buyers. For example, a 2003 report from the University of Wisconsin–Stevens Point Center for Land Use Education found building on lake shorelines proliferating. From 1960 to 1995, the number of dwellings along 235 northern Wisconsin lakes increased by an average of 216 percent. At the same time, people built bigger homes, while lot sizes did not increase in proportion. "If present development rates persist," the report stated, "all undeveloped lakes not in public ownership could be developed by 2020."[7]

While that scenario has not literally come true, the development of lake shorelines has been significant, and its effects appear in current prices of undeveloped lakefront lots in many areas. For example, says Zastrow, the lower prices for lots late in the past decade in part reflect that the most desirable and thus most valuable lots have largely been built upon. Those remaining are likely to be on less desirable lakes, to have construction constraints, to have building sites with distant setbacks from the water, or to have weedy and mucky shorelines known as "fisherman's frontage." "The best of the best lots are long gone," he says.

2000

Sunday Lake Cottage. Perfect getaway spot on a picturesque lake. Very well maintained smaller cottage ready for immediate use in a great location. Nice place for kids to play and easily used 4 seasons. *$129,000*.

Nonetheless, vacant lots have continued to sell, while the traditional seasonal family cabin has steadily faded from lakescapes, for a variety of reasons. One is that their owners are aging. A 2016 survey by Minnesota Lakes and Rivers Advocates found that the average age of lake home and cabin owners in that state was sixty-eight; it was sixty-two in 2005 and fifty-eight in 1999.[8] For many, this means the time is approaching when they will need or want to sell. When that happens, whether sold, willed to family members, or offered on the market, there is no guarantee the little cabin will remain. "What is fairly common now is that aged housing stock is bought, torn down, and because the value of the property is so high, someone will build a correspondingly high-priced home," observed Tim Houle, administrator in Crow Wing County in northern Minnesota's Brainerd Lakes region. He no longer refers to that area as cabin country: "I don't think that is what we have, and we haven't had it for 20 or 30 years. These are lake homes."[9]

Sandra Ebben sees the same trend in northern Wisconsin: "Even if it's a little house, people will buy it basically for the lake frontage. Maybe they'll use it for five years as it is and then, depending on what the structure is like, either tear it down or expand it."

Property taxes also come into play, notes Jeff Forester, executive director of Minnesota Lakes and Rivers Advocates. Contrary to what many believe, cabin owners are not universally wealthy: his group's 2016 survey found the average household income of Minnesota seasonal property owners was $58,000. As property values rise, so do taxes, to the point where some owners of modest means no longer can afford to pay them. They are then forced to sell or, if they own larger parcels, encouraged to subdivide. Forester's family owns property with a rustic cabin and extensive water frontage on an island in Minnesota's Vermillion Lake, going back to 1918. "In 1995, the property taxes were about three hundred dollars a year," Forester says. "Today [2020], the land has been subdivided, so that where there was one property owner, now there are five, and the property taxes cumulatively are pushing twenty thousand dollars." His family still owns one of the parcels.

Small seasonal cabins like this one are fading from the lake country landscape.
(Ted Rulseh)

Large year-round lake homes often take the place of family summer cottages.
(Ted Rulseh)

In addition, a generational change is part of the picture. Forester observes: "People under the age of, say, thirty-five to forty, the Millennials, have huge college debts. They're not buying homes, they didn't get married in their early twenties, and by and large they're not having 2.5 kids. And they don't want to own things. They're much more interested in VRBO or Airbnb. If they want to have a nice family time on a lake, they'll rent a cabin for a week. They don't have to do any maintenance, and in the long run it's way cheaper than buying something."

2005

Bass Lake Cottage. Located on a level lot, 99 feet of sandy shoreline and southwest exposure, giving you hours of sunshine and beautiful North-woods sunsets. Enjoy the abundant wildlife and summers at the lake. $179,000.

Health-care costs are another factor forcing families to sell treasured cab-ins, Forester believes. "Since the 2000s, we see many cases where mom and dad get sick and end up in assisted living or in a hospital or nursing home," he says. "The kids sell the parents' house in town. They burn through all that money for the medical bills, and then the last thing to go is the cabin. There's no money left after end-of-life costs. The health-care system depletes every asset they've got. The little two-bedroom cabin gets knocked down, and a big home gets built."

Finally, there are practical difficulties with keeping traditional lake cab-ins within families. When the parents are ready to sell or transfer the prop-erty, various complications interfere. For example, "There are four siblings; two can afford to buy the cabin, two cannot," says Zastrow. "Or three live in Illinois, while one lives in California and can never use the cabin." Dis-putes and conflicts among family members arise, and the best remedy is simply to sell the property to a new owner, who will most likely knock the cabin down. "It might even be a halfway decent house, but it still represents an underimprovement for the lot value and the quality of the lake," Zastrow observes. "The days of these properties passing to the next generation are largely gone. Little cabins don't exist much anymore." Exceptions include pockets of cabins on smaller and otherwise less prestigious lakes.

The net results of all these trends are more and larger homes per mile of lake shoreline, more structures, more and bigger boats and other water-craft, more septic systems, and, in general, more threats to scenic values and more stresses on lake ecosystems.

CHAPTER 4

One Water

We tend to think of lakes, streams, wetlands, and groundwater as separate entities. In reality, they are all part of one interconnected system, driven by the water cycle of rainfall and evaporation.

Paul McGinley, a water quality specialist with the University of Wisconsin Extension Center for Watershed Science and Education, explained this with clarity in a presentation at the Wisconsin Lakes Convention in April 2019. He starts with the water that falls as rain or snow. His home state of Wisconsin gets on average thirty-two inches of precipitation annually. Of that, about twenty-two inches evaporates or is taken up by the roots of trees and plants, which transpire most of it back to the air through their leaves. Much of the rest, about ten inches on average (a range of about eight to fourteen inches) may enter the groundwater system.

Buried Treasure

There is far more water below the surface in the state than in all the lakes and streams combined. It has been estimated that the groundwater beneath Wisconsin would be enough to cover the entire state to a depth of one hundred feet.[1] In a bigger picture, scientists estimate that groundwater in the United States comprises at least thirty-three thousand trillion gallons, about as much as the Mississippi River has emptied into the Gulf of Mexico over the past two hundred years.[2] This water doesn't exist as an underground lake or river. Instead, it fills the spaces between the particles of sand and gravel. If you take a glass, fill it with sand, and then pour in water until the sand is saturated, that is how groundwater exists far below our feet. It does flow, very slowly, from higher to lower elevations, obeying the law of gravity. "It is possible for water to flow from lower to higher elevations when hydraulic pressure drives the water upward," says McGinley. "This is most conspicuous in artesian wells."

The figure below illustrates how the groundwater flows out to surface waters. Rain falls on the landscape, and as the water percolates through the soil, it mounds up slightly at the higher elevations. The top of the groundwater reservoir is known as the water table. The water gradually migrates, following the slope of the landscape, until it finds an outlet in a lake or stream. It can take years or decades for a drop of water to travel from where it fell on the ground to a body of water. The slow, steady flow of groundwater is the reason streams can run all year even during long spells with no rain. Meanwhile, says McGinley, lakes can be thought of as places where a depression in the landscape causes the water table to be exposed. That is, looking out across your favorite lake, you are in effect seeing the surface of the local groundwater. Of course, not every year brings exactly the same amounts of rain and snow. In dry years, there is less new water to recharge the groundwater system, and the water table will fall. In wet years, the opposite happens.

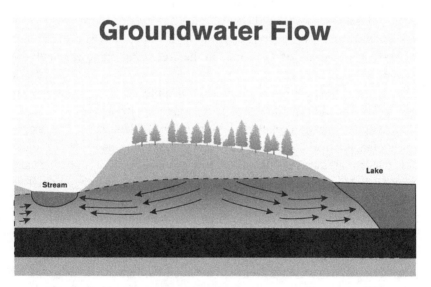

Groundwater Flow

Groundwater flows from higher to lower ground and ultimately enters a lake or stream. (Eric Roell; adapted from "Where Did That Water Come From? Precipitation, Groundwater, Streams and Lakes," presented by Paul McGinley, Wisconsin Lakes and Rivers Convention, April 2020)

All lakes interact with the system of rainfall, groundwater, and streams. There are three main types of lakes: drainage, seepage, and spring. On drainage lakes, a stream brings water in, and a stream takes water out. The water level in these lakes tends to stay fairly constant. Think of a bowl into which you run a slow flow of water from a tap: When the level rises to the top of the bowl, the water spills over. An equal amount comes in and goes out; the level is self-regulating. Drainage lakes usually also have connections to the groundwater.

Seepage lakes have no stream flowing in or out. Their water comes from rainfall and runoff, supplemented by the flow of groundwater. Because the water table changes in dry and wet periods, the water levels in these lakes (sometimes called spring-fed lakes) tend to be cyclical. Spring lakes get their water in the same way as seepage lakes, but they have a stream outlet.

Understanding Watersheds

Whatever its type, no lake is an entity on its own. Every lake is part of a landscape feature called a watershed. To understand the concept of watershed, it helps to picture a large, shallow bowl. Water falling anywhere on the bowl's surface will run toward the low point at the center. In a landscape, hills and slopes form the sides of the bowl; water drains down to the lakes and streams at the bottom. Scientifically speaking, a watershed includes all the lakes, streams, and wetlands in the area being drained, and all the groundwater underneath.

A lake's watershed can be quite large or quite small. The waters near where I live provide an example. About a mile from my house lies forty-six-acre Muskie Lake, at a relatively high point in the landscape. The lake is spring fed; its watershed covers only the surrounding wooded land and the few residential properties that drain into it. A creek connects Muskie Lake with Little Muskie Lake (nine acres), which in turn feeds thirty-eight-acre Seed Lake. From there a creek winds through swampy woodlands to Birch Lake (180 acres), where I live. The Birch Lake watershed includes its immediate surroundings, along with the land that drains water into the creek and the three smaller lakes upstream. And so it goes, downstream to Big Bearskin Lake, Little Bearskin Lake, Bearskin Creek, the Tomahawk River, and on to the Wisconsin River, the Mississippi, and the Gulf of Mexico. Each body of water farther downstream has a progressively larger watershed.

This interconnection has implications for the condition of the lakes as it relates to how property owners develop their land and the good or bad practices they follow. Residents of Muskie Lake largely can control their own destiny. Their water quality depends on what they do; there are no major impacts from upstream. Big Bearskin Lake, on the other hand, could see effects not just from activity on its shores but from land use practices on Muskie, Little Muskie, Seed, and Birch Lakes. Even in such cases, because the land in most northern lakes' watersheds consists of natural forest and wetland, the impacts from upstream may be mild. Lake residents can then band together to protect or improve water quality through sound property management.

The picture changes where the watershed includes farms that contribute nutrients from fertilizer, or a city or village that releases stormwater with all manner of pollutants, from nutrients to oil and gasoline to sediment, leaves, and grass clippings. Here, improving a lake water-quality problem means a long, challenging, and costly process of gaining cooperation from diverse landowners upstream. This typically includes forming an organization of community leaders, farmers, educators, municipal officials, environmental advocates, and others to develop a watershed management plan.

The key point is that water quality starts on land. Lake protection or improvement always involves a community—whether of property owners on a single lake or chain, or of multiple interests covering a wide geography. All the waters are connected, and in working to protect those waters, so are all of us.

CHAPTER 5

The Trouble with Phosphorus

In twenty-four years living on Michigan's Six Mile Lake, Cherie Hogan had seen nothing like it. In late August 2019, blue-green algae bloomed on the lake, a peaceful place surrounded by woods, a lake on which loons each year raise healthy chicks. "This year was a strange year," Hogan told Michigan Public Radio. "We had the bloom; we've never had it here before. Then all of a sudden, the conditions just became right and it was in the lake."[1]

Six Mile Lake, 370 acres with a thirty-one-foot maximum depth and 8.7 miles of shoreline, lies in Antrim County in the northern part of Michigan's Lower Peninsula, where lake algae blooms are uncommon. Its phosphorus levels generally are low. It has a diverse, healthy, and stable array of native aquatic plants, along with invasive Eurasian water milfoil, which the Six Mile Lake Association is working to control.[2]

BreAnne Grabill, manager of Northern Regional Lakes for PLM Lake and Land Management Corporation, verified the presence of blue-green algae. "It was a large bloom," Grabill noted in an interview with a local paper, "one of the largest blooms I've ever seen in northern Michigan."[3] It involved the entire lake, neon blue-green splotches drifting everywhere and accumulating in uniform slicks where breezes pushed it up against shorelines. The bloom persisted for two weeks.

"Six Mile Lake has always had some filamentous algae, globs of what some people call Ghostbuster slime, but it was an aesthetic issue; it was very minimal," Grabill says. "Right around the Fourth of July we were doing a plant survey on the lake, and we noticed some small blooms of planktonic algae starting in different areas of the lake. Within a matter of days it bloomed lakewide. Northern Michigan has a lot of really beautiful water bodies, and Six Mile Lake is definitely one of them. It really surprised me when a technician brought back a sample. I said, 'You found that there?' I never expected it. It was really shocking."

⁓⁓⁓

The root of the problem was an excess of nutrients, which many experts regard as the most universal and pervasive threat to our lakes. Nutrient pollution has a variety of causes, and many relate directly to the ways we live on our shorelines. Nutrients in themselves are not pollutants—they feed the algae and aquatic plants that are essential to the lake food chains and the lake ecosystem. It's only when overabundant that they create problems.

All plants, on land or in the water, rooted in soil or sediment or free-floating in a lake, require a variety of elements. The most basic are carbon, hydrogen, and oxygen, which make up the cellulose, sugars, and starches that form plant tissues—roots, stem, leaves, flowers, fruit, seeds.

Next are the three primary nutrients: nitrogen (chemical symbol N), phosphorus (P), and potassium (K). If you buy fertilizer, you'll see printed on the bag an NPK analysis that shows the percentage of these nutrients in the formula. For example, a lawn fertilizer with an NPK analysis of 10–0–5 contains 10 percent nitrogen, zero phosphorus, and 5 percent potassium. Most lawn fertilizers today are free of phosphorus, for reasons that will soon become clear.

These primary nutrients perform specific functions as plants and algae live and grow. For example, they are components of the basic plant tissues and structure. They are part of the DNA in plant cells. They help enable photosynthesis, by which plants use the energy of sunlight to make food from water and carbon dioxide, and they act as catalysts for biochemical reactions. Plants also need lesser amounts of calcium, magnesium, sulfur, copper, iron, and other nutrients, but N, P, and K are the most essential. If even one of these is deficient, plant growth and reproduction are constrained or impossible.

Effects of Overfeeding

In our lakes, the main nutrients of concern are nitrogen and phosphorus. These, when in excess, accelerate a lake aging process called eutrophication. From the perspective of plant life, eutrophication is basically too much of a good thing. It's caused by a nutrient overabundance that sends plants

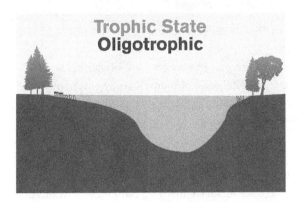

Trophic State
Oligotrophic

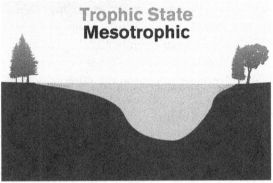

Trophic State
Mesotrophic

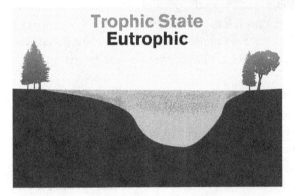

Trophic State
Eutrophic

One way in which scientists classify lakes is by nutrient status—in essence where they fall in the process of eutrophication. On one end of the spectrum are oligotrophic lakes, low in nutrients with sparse plant life, little algae, and usually very clear water. On the other end are eutrophic lakes, rich in nutrients, with heavy plant growth, and often subject to algae blooms. In between are mesotrophic lakes, with moderate amounts of nutrients, plants, and algae. The uncontrolled addition of the nutrient phosphorus over time can cause lakes to become eutrophic more rapidly than they would through natural processes, and so more subject to prolific growth of aquatic plants and nuisance blooms of blue-green algae. (Eric Roell) and algae into a frenzy of feeding and growth. Our lakes naturally accumulate nitrogen and phosphorus as they age. Along the way, plants and algae reproduce, grow, die, and decompose. Ever so slowly, the lakes fill in with dead plant matter, other organic material, and sediment that washes in from the land. If lakes are left alone, this aging process unfolds over hundreds or thousands of years. But a rapid increase in nutrients introduced by humans can speed up the process significantly.

The lake can be overrun by thick masses of aquatic plants that grow up to the surface and impede boating and swimming. Algae can multiply out of control (bloom), turning the water green. Any type of algae can bloom, including the largely benign and beneficial green algae. Most troublesome are explosions of blue-green algae, really not algae at all but microorganisms called cyanobacteria. These events, called harmful algal blooms, create the conditions Six Mile Lake residents observed in 2019.

The impacts on lake health can be severe. As the days go by and the cyanobacteria die, they decompose, using up the oxygen in the water. In severe cases, oxygen becomes so depleted that fish can suffocate and die. Blooms also can reduce populations of green algae that zooplankton such as *Daphnia* (water fleas) need to survive. These tiny creatures are a key food source for small and developing fish, and as the *Daphnia* decline, the lake's food web is disrupted. In short, the lake ecosystem is distressed. And that is not the only problem.

On a morning in July 2014, Jean Roach's phone started ringing. "What is going on?" the callers asked. "What is happening on the lake?"

Roach, then president of the Pelican Lake Property Owners Association, couldn't see a big problem from her shoreline on Musky Bay. But across the lake, on North Bay, association board member and science specialist Dave Hardt observed a dense, unsightly film on the water, especially amid the bulrushes along his shoreline. He immediately suspected a blue-green algae bloom.

It wasn't unusual for green algae to give the lake a pea-soup color in August, but this bloom was different. "It was ugly," Hardt recalls. "It was a scummy, oily-looking film, a weird bluish-green with some yellow. It looked like somebody spilled paint in the water. It stunk to high heaven. I have a room in my house, in the back away from the lake. I was taking a nap in there one day; I had the windows open, and I almost couldn't sleep because it smelled so bad."

Such an event in the middle of summer was cause for deep concern. Pelican Lake, the largest lake in northern Wisconsin's Oneida County at 3,545 acres, is a prime tourist attraction in the area. Everyone—owners of resorts, bait shops, and bars, lake property owners, visiting boaters, and anglers—had the same question: What is wrong with the lake?

~~~~~

The Limiting Nutrient

Excess nutrients in lakes come from sources that include polluted runoff from farmland upstream and from improperly designed, poorly maintained, or failing septic systems. Other sources include stormwater carrying all manner of dirt, chemicals, and debris from urban streets; overapplication of lawn fertilizers; and rainfall runoff from lakefront lots carrying leaves, pet waste, and other nutrient-rich materials.

The nutrient of greatest concern for most inland lakes is phosphorus. By its nature, phosphorus is essential; it supports the production of plant roots, flowers, and fruits and gives strength to stems and stalks. But in excess, it is destructive to lake ecosystems. Phosphorus is known as the limiting nutrient in lakes—among N, P, and K, it is the essential nutrient typically in shortest supply. Our lakes contain enough nitrogen, potassium, and other nutrients to support plants and algae. Nitrogen may be the limiting nutrient in some lake ecosystems, but in the inland lakes of the Upper Midwest, it is usually a shortage of phosphorus that keeps a lid on prolific growth.

An analogy from *Lake Tides*, a newsletter published by the University of Wisconsin Extension, explains the concept of a limiting nutrient: "A pound cake takes a pound of flour, a pound of butter, a pound of sugar and four eggs. If you have ten pounds of flour, butter, and sugar, but only four eggs, you can only bake one pound cake. The eggs are the limiting factor to baking more."[4] In a lake, phosphorus is like those eggs. The issue is not that plants and algae need a great deal of phosphorus. The issue is that in most lakes, phosphorus is comparatively scarce. But when more phosphorus is added to the water, the algae have a feast and reproduce with abandon. It takes forty pounds of carbon, seven pounds of nitrogen, and just one pound of phosphorus to grow five hundred pounds of algae (wet weight). Because ample carbon and nitrogen already exist in the water, just one additional pound of phosphorus can yield another five hundred pounds of algae. This is often called the Algae Adage.[5]

On seeing the blue-green algae bloom in Pelican Lake, Roach and Hardt worried about the risks to lake residents' and visitors' health. Some species of cyanobacteria produce toxins that can sicken people who swim in or drink the water, or even inhale spray from motor-boating or waterskiing. The effects can include skin rashes, headaches, vomiting, and diarrhea, and more serious problems like seizures, paralysis, liver damage, and hemorrhaging. Children and pets are especially vulnerable. Dogs can die from eating blue-green algae washed up on shore, or from licking their fur after swimming during a bloom. A dog sickened by these toxins may die before its owner can drive it to a veterinarian's office. Cyanobacteria toxins also have been known to kill waterfowl.

"Once testing had confirmed that it was blue-green algae, we needed first of all to let people know the dangers," says Roach. "Pelican is a lake where a lot of people from the area come to fish. We have a lot of weekend traffic. We made informational pieces, laminated them, and gave them to our boat landing monitors. They used them to tell people about the risks of swimming in it, the dangers of letting their dog swim in it." A local TV station learned of the bloom and filmed a news report in which Roach and Hardt described the hazards; it included footage of the bloom taken from Hardt's property. "My pier was right in the middle of the worst of it," Hardt recalls. Some business owners were not happy with the publicity, "but we persisted because it was the right thing to do," Roach says. "Many people were very appreciative, particularly people with children and pets."

Sources of Phosphorus

Most of the phosphorus added to our lakes comes from runoff and erosion from the surrounding land and from feeder streams. Some of that happens naturally and can't be prevented—for example, rainwater draining off the land carries phosphorus from leaves and other dead plant material and generally some amount of soil. But things people do on their properties can feed a great deal more phosphorus into the water. Lakes fed by streams that pass through farmland are the most vulnerable, especially if the farmers apply too much phosphorus fertilizer or spread manure without taking steps to keep it from washing off the land. Some farmers routinely apply phosphorus fertilizer to land already rich enough in that nutrient.

Phosphorus exists naturally in soil and takes different forms. Most of the phosphorus lies in the topsoil. The majority is tightly bonded to soil particles so that it cannot be used by plants; the rest is dissolved in water, can easily travel with runoff, and, once in the lake, is readily available to plants and algae. As rainfall or snowmelt passes over the ground, it picks up both forms of phosphorus. The amount largely depends on the volume of runoff water and the speed at which it flows. Slow-flowing water mainly carries dissolved phosphorus. Fast-flowing water, especially if passing over soil not anchored by plant roots, can cause erosion, and most of the phosphorus it carries is attached to soil particles or embedded in bits of dead plant material.

The dissolved phosphorus can immediately boost the growth of rooted water plants and algae. The soil and plant matter particles eventually settle to the lake bottom, acting like a storehouse of phosphorus. From there, under the right conditions, bacteria can convert the phosphorus to the soluble form and release it into the water, giving blue-green algae an environment rich in a critical nutrient.

Harmful Blooms

Like true algae, cyanobacteria make their own food through photosynthesis. Unlike true algae, they are not a major food source for lake creatures. They take their common name from their characteristic color, but they can also manifest shades of green, blue, brown, or reddish-purple as they die, decompose, and release their pigments. In northern latitudes, algae blooms tend to begin in mid-June and can occur until late September. Under certain conditions, the blue-greens have competitive advantages over most other algae. For example, they thrive at warmer temperatures and can prosper even under relatively low light and in turbid water. Blooms can appear quite suddenly—even overnight. Some cyanobacteria have evolved the ability to control their buoyancy; during the day they can suspend below the water's surface and multiply there, taking advantage of sunlight. After dark, they float to the surface, so that in the morning a noxious scum or floating mat appears. A persistent wind can push the algae on the surface and concentrate it along a shoreline, creating an ugly, bad-smelling, toxic mess. There it can stay until it is scattered by wind and waves, or until lower sunlight and cooler days allow it to dissipate. Every year, state and county health departments are forced to issue warnings about toxic algal blooms in lakes.

Not all blue-green blooms can be traced directly to shoreline owners' practices. Some lakes are naturally high in phosphorus from sources such as surrounding woodlands and marshes. These lakes can take in fairly low additions of nutrients and yet still experience blooms. Furthermore, a bloom generally does not spring from a sudden, large influx of nutrients. Blue-green algae are always in the water, just waiting for conditions to be right.

Many factors determine how much phosphorus lakes can assimilate and how vulnerable a given lake will be to blue-green algae blooms. One major factor is retention time—how long it takes for water to cycle through a lake, according to Bob Martini, retired statewide river protection coordinator with the Wisconsin Department of Natural Resources and president of the Oneida County Lakes and Rivers Association. For example, in a lake with a stream flowing through, it may take only a few years to flush excess phosphorus out of the system, while in a lake that receives water only from rainfall and a connection to groundwater, excess phosphorus may persist for decades as the water turns over very slowly.

On Pelican Lake, the association leadership believed that failing septic systems added phosphorus to the lake and contributed to the bloom. "There were a lot of cabins built on wetlands in the 1930s and 1940s," Roach says. "Some of them have the original septic systems in operation today. The county now requires a septic system check when properties change hands, but many of them on Pelican are passed on to family members, and those transfers do not trigger a check." A consulting firm for the lake association suspected a process called internal nutrient loading as a cause of the bloom. In that process, the water above the sediment becomes devoid of oxygen; bacterial activity then releases phosphorus normally tied up in the lake bottom. Wind action can then stir up that phosphorus-rich water and bring it to the surface, where algae take it in and multiply. Pelican Lake also takes in natural phosphorus from its large adjoining wetlands. The cause of the bloom on Six Mile Lake remains uncertain, although Grabill believes that high water levels in recent years, releasing more phosphorus than normal from the newly submerged land, could have been a factor. In either case, the lesson is the same: more phosphorus in the lake makes blooms more likely and potentially more severe.

Once a pattern of blooms takes hold in a lake, no easy or low-cost treatment can reverse it. Blooms can disappear in days or persist for weeks; sunlight and warm, calm days tend to make them last longer. To prevent blue-green algae blooms, some lake associations treat the water with aluminum sulfate (alum), which removes phosphorus by a process of precipitation. The alum forms particles called floc that slowly settle to the lake bottom, creating a barrier against release of phosphorus from the sediment. Alum treatment is proven safe and effective, but it is also costly and may need repeating at intervals of several years; it essentially treats the symptom and not the root cause of blooms.

The 2014 Pelican Lake algae bloom persisted for about ten days—days that were hot and almost windless. "It's normally quite a windy lake," Hardt says. "We can get a lot of whitecaps. Once the winds picked up again, it stirred up the water, and the ugly color and the stink went away." As of 2021, there had been no other major blue-green blooms in the lake.

As for Six Mile Lake, the lake association used chelated copper and copper sulfate to treat areas where the blue-green algae had accumulated along beaches and developed shorelines. That is only a short-term solution used in strategic spots; it is generally not a practical remedy for a lakewide bloom, and in fact excessive levels of copper can be toxic to lake life, especially invertebrates. After the bloom, lake property owners were left to wonder whether it would recur in future years. "Time will tell," says Grabill. "It could have been a fluke where the circumstances all met together. By no means should Six Mile Lake go down as one that has massive planktonic blooms every year. There is nothing to indicate that this will be a routine situation."

Still, the Six Mile Lake experience is not unique in its general area. Dave Edwards, former monitoring and research director with Tip of the Mitt Watershed Council, told Michigan Radio that harmful algae blooms are "relatively new" in his organization's territory, which includes four counties in the northern part of Michigan's lower peninsula.[6]

The Longer Term

Jennifer Buchanan, associate director with the Tip of the Mitt Watershed Council, says it's not entirely clear whether harmful blooms are becoming more prevalent. "I would say, without qualifying it too much, that yes, we're expecting more and we think they're happening more, but documentation has been incomplete. Sometimes when algal blooms happen, there's an assumption that they either are or are not harmful algal blooms." Her council is stepping up communication with the Michigan Department of Environment, Great Lakes, and Energy to test blooms and conduct outreach to make sure that lake residents don't wade into a suspect bloom to collect samples. "If they see any algal bloom at all," she says, "we're directing them either to contact us so on their behalf we can contact the state, or to go directly to the state, rather than try to collect a sample and bring it in. People should stay out of the water if there are any characteristics at all that indicate an algal bloom could be harmful."

On a deeper level, Bob Martini, of the Oneida County Lakes and Rivers Association, worries about the long-term future of lakes if phosphorus inputs are not curtailed. He says,

> We already see people coming to the Department of Natural Resources and saying, "I've never noticed this much algae in my lake before. I've never noticed so many plants growing. Why don't you fix it? Why don't you bring it back to the way it was when my dad bought our place forty years ago?" Well, after a certain point, you can't go back. Once you reach a tipping point, those nutrients are cycled internally and you're stuck with that level of nutrients for a long, long time. Wisconsin did a great deal of research in the 1960s, 1970s, and 1980s on how to get rid of phosphorus in lakes. The costs were huge, and the conclusion was that it isn't feasible. You can remove some phosphorus, but you can't bring a lake back to its original condition.

Lake scientists agree that the best medicine against phosphorus accumulation and blue-green blooms is prevention: limiting new additions of phosphorus to the water. Steve Carpenter, former director of the Center for Limnology at the University of Wisconsin–Madison, has observed that controlling phosphorus in a lake can be like walking on a treadmill while someone constantly dials up the speed.[7]

CHAPTER 6

Changing Lakescapes

Joe Davidson lived for fishing in northern Michigan. In 1960 he bought a lot with a hundred feet of frontage on the 150-acre Sunset Lake for himself and his fishing pals to use on weekends, and for his wife, Lucille, and their three kids to enjoy for a week or two each summer.

At the time, the property, one of several for sale on the lake, was fully wooded, a mix of hemlocks, oaks, maples, and half a dozen towering white pines on a moderate slope from the access road down to the water. Rainfall hitting the leaves and needles dripped softly to the earth, where it soaked quickly into the leaf litter and sandy glacial soil. In heavy rains, the roots of trees and understory plants held the soil in place. Some runoff reached the lake, carrying with it organic matter from leaf fragments, and at times a little sediment, but not enough of any of it to enrich the water greatly. In other words, the lot had about the same low impact on the lake as before the land was logged decades earlier.

At first Joe just cleared away a few trees to make a place for himself and his friends to pitch tents during their fishing escapes from the Detroit suburbs. Then he added a picnic table and a fire pit surrounded by crude benches made from half logs. After a few years, Joe parked a camping trailer at the top of the slope. By the late 1960s, he had scraped enough money together to drill a well and build his cabin, a modest six hundred square feet, with white-gas lamps, propane-fueled refrigerator and stove, and a hand pump beside the kitchen sink to bring up ice-cold, slightly iron-tasting water. An outhouse took the place of a bathroom. Over the years, Joe added electricity, plumbing for the kitchen, and a log-sided shed to store firewood and a few tools. The property still existed in harmony with the lake, the woods mostly intact, just a relatively few square feet of cabin roof catching the rain and increasing the flow of water down the hill.

Then 2005 brought a big change. Joe and Lucille were in their late seventies, the kids grown up and living in different states, with children of their own. The two-hundred-mile trips to and from the city were burdensome now, and so was the upkeep of the cabin and property. Chores like splitting and stacking firewood, fixing screen doors, patching the roof, and blowing out the water pipes for winter weren't fun anymore. Climbing the thirty-five steps between the cabin and water had become a struggle.

They wanted to keep the cabin in the family, but their kids weren't interested, nor were other younger people among their extended relations. There was nothing to do but put the property on the market. Lucille cried a little the day they signed the listing contract with the real estate agent and watched her push the For Sale signs into the ground, one near the edge of the road, the other down on the lakefront. Lucille cried again, and Joe did, too, when the sale became final and they signed the closing papers. Unable to bring themselves to attend the closing at the bank, they never met the new owners. The check they received by mail, many times what they had paid for the lot, brought only a little solace.

Now more change was in store for the lot on Sunset Lake. In early June, a month after the closing, the new owners hired a contractor to make room for a house. The contractor brought in a yellow-gold John Deere excavator and, with several masterful strokes of the bucket, ripped the little cabin to pieces. Another contractor cut down trees to prepare for a house with a footprint more than twice the size of the cabin, with fifteen feet of clearance all around the space where the structure would stand. More trees came down to make room for a septic system, and still more to open a corridor and create an unobstructed view of the lake from what would be the family room and the second-floor screen porch and deck. The house went up that fall, with a detached garage big enough for a car, a pickup truck, two snowmobiles, and winter storage for two boats. A wide asphalt driveway connected the garage with the road. On the cleared land, the owners planted a lawn, which they mowed regularly; it reached nearly all the way to the lake, ending at a narrow strip of sand beach. The next summer saw a boathouse built down along the water, with a screened-in party room above it.

Whether or not the owners knew it, the half-acre lot now affected the lake profoundly and far differently than in all the years before. Just a few scattered trees stood on the lawn; only the property's fringes and a patch on each side of the boathouse remained wooded. Boaters and anglers on the lake no longer saw Northwoods scenery but the boathouse door and an expanse of grass more at home on a lot in the city. The strip of near-shore land, the area richest in wildlife habitat, was largely stripped of trees and natural vegetation.

Atop the slope, rain hitting the road and the asphalt sluiced down the driveway, rushed between the house and garage, and swept across the lawn toward the water. In major storms, the house and garage roofs funneled rainfall into gutters; water gushed from the downspouts and flooded down the slope, striking the boathouse walls and carving soil away. Runoff from the boathouse and surroundings drained directly into the lake. All across the property, raindrops now splatted full force on the close-cropped grass, kicking up soil particles and bits of dead matter, which the flow of water carried into the lake. That water also carried oil, gasoline, antifreeze, salt, and particles of tire rubber from the driveway; fertilizer and traces of weed killer from the grass; and nutrients from dog droppings scattered around the yard. Heavy rains left the near-shore water roiled light-brown with sediment.

All this unfolded in just a few years after Joe and Lucille sold their property. When they paid a visit to their old getaway on a nostalgic weekend trip, what they saw moved them, once more, to tears.

Joe and Lucille are fictional, and so is the Sunset Lake described here, but stories like these have played out for decades in very real ways across the northern lake country. The transformation of just one small woodland cabin to a large home with an urban-style landscape would have little effect on a lake. But twenty-five, or fifty, or a hundred lots similarly changed can have substantial impacts—on scenery, wildlife habitat, water quality, and the value of lake properties. In fact, excessive development of shorelines can dramatically alter a lake's character, destroying the mystique that once attracted property owners and visitors. Patrick Goggin, a lake specialist with the University of Wisconsin Extension, observes that the number one stressor of lakes cited in the EPA National Lakes Assessment in 2007 and 2012 was the loss of natural shoreland habitat.

In many cases the impacts are readily apparent even to casual observers. Heidi Shaffer, soil erosion officer with the Antrim Conservation District in northwest Lower Michigan's Antrim County, has seen the effects on the Elk River Chain of Lakes, a seventy-five-mile-long waterway consisting of fourteen lakes and connecting rivers. Says Shaffer,

I have been in this position for twenty years, and I have seen a significant change in the quality of the lakes in our region. We have Torch Lake, which is touted as one of the most beautiful lakes in the country. When I came here in 1998, the lake bottom was clear and beautiful. It's an oligo-trophic lake and its water quality still is good by downstate standards. Today it has golden brown algae on the bottom, and we have issues with blue-green algae in our chain. I personally think it's the result of the sub-urbanization that is going on. People come here from urban areas all over the country where they have perfectly manicured lawns, and they want the same thing here. They remove the natural vegetation and plant grass lawns all the way down to the lake. And I say, "You came up here because of the beauty and the clean water, and now you're replicating what you left behind?" It really doesn't make sense. The water quality and the beauty of this area are directly tied to the natural vegetation around our lakes: trees and shrubs and native plants are critical to preserving the Up North aesthetic and the water quality we all enjoy.

Goggin, in describing the qualities of a "healthy, robust, intact lake-shore," defines distinct zones that progress from the near-shore water up to higher ground. Some distance out from shore are submersed plants such as coontail, assorted pondweeds, and water celery. A little closer in are floating-leaf plants like water lily, spatterdock, and watershield. Near the shoreline in a foot or two of water are emergent plants (rooted in the bottom but standing above the waterline), such as bulrushes (sometimes called reeds) and the purple-flowered pickerelweed with arrowhead-shaped leaves. "These plants provide food and cover for wildlife," Goggin says. "The little baby fish can live in there, along with zooplankton and phytoplankton that provide them with food." The plants give oxygen back to the system, create cooling shade, and provide vertical structure on which creatures such as dragon-fly and damselfly nymphs can climb. They also help protect the shoreline against erosion: "When waves come in, those plants are bobbing up and down, helping to dissipate some of that energy."

Another component of in-water habitat consists of tree branches, logs, and stumps lying on the bottom. "In a natural lake where fallen wood is left in place over time, as many as four to five hundred pieces of wood or logs can be found along each mile of shoreline," Goggin says. This wood provides more fish and wildlife habitat and adds protection for the shoreline against wave action and springtime ice pushes. "It also helps quiet the water down a little bit to give aquatic plants a chance to get established," he adds. A natural lake bottom in itself, perhaps a mix of rocks, gravel, sand, and muck, also supports a variety of life, providing habitat for clams, mussels, and aquatic insects, and nesting areas for spawning fish.

Vegetation Zones

On the land nearest the shore are three layers of vegetation. Explains Goggin,

> Starting down on the ground we have wildflowers, sedges, rushes, and ferns growing. Next we have a mid-layer of shrubs, small trees, and taller flowering plants. Then there's the canopy. Most of our lake edges in the north have a forest environment; on some lakes we're lucky enough to have a super-canopy, the white pines popping up over the tops of the deciduous trees. Those are used by eagles and ospreys. This suite of native vegetation taken together provides a buffer of food and cover for wildlife. This shoreland area is home to a great diversity of life—bacteria, fungi, insects, snails, worms, frogs, toads, snakes, turtles. In fact, 80 to 90 percent of all life is born, raised, or fed in the near-shore zone.

But probably the most important function of the vegetated shoreland is in slowing down the flow of rainwater and snowmelt, spreading it out and letting it infiltrate into the ground, instead of carrying sediment and nutrients into the water. The plant roots hold the soil, and the heavy layer of duff on the forest floor—a mix of leaves, evergreen needles, and twigs—supports a variety of plants and also acts like a sponge: "If you pour a jug of water into that duff layer, it goes away, like that!" says Goggin. "Rainfall onto it just gets soaked up. This is what an intact, healthy lakeshore looks like."

Residential development in many places has markedly changed the nature of lake shorelines, and one significant factor has been the demise of small lake cabins, according to Goggin.

> The typical three-season cottage or cabin often had a grass access corridor to the water's edge about twenty feet wide. The cabin might be twenty-five

by twenty-eight feet, or about seven hundred square feet, with a gravel driveway and a natural shoreline buffer of thirty-five feet or more. The impact of this building pattern on the lake during the April to October growing season could be up to a thousand cubic feet of runoff, 0.03 pounds of phosphorus, and twenty pounds of sediment. That's how we used to enjoy lake life.

But beginning in the 1940s and continuing today, we changed those cabins into three-thousand- to four-thousand-square-foot second homes, with hard roofs, a garage, an outbuilding, and a driveway that's black-topped. All of a sudden we have increased the hard surfaces on that one-hundred-by-two-hundred-foot lot, and during the rebuilding we don't take into account the increased water runoff and sediment movement. We can see runoff to the lake jump to five thousand cubic feet, 0.20 pounds of phosphorus get to the lake, and up to ninety pounds of sediment as well.[1]

Regulating Runoff

Today, preventing those additions is largely a matter of controlling the water that falls on the land. An inch of rain falling on one acre delivers 27,152 gallons, about enough to fill a backyard swimming pool thirty feet

Some large homes are landscaped with lawns leading nearly or all the way to the shoreline. (Ted Rulseh)

long, twenty feet wide, and six feet deep. A landscape of native plants is well adapted to absorb that water. The roots sink anywhere from a foot to several feet into the ground. "About a third of that root structure dies back every year, creating little channels for the water to infiltrate," Goggin says.

If that kind of landscape is like a sponge, then a lawn is like a sponge already partly compressed; it absorbs significantly less water. Even a lush turf will send roots only about four to six inches deep. "And we can't grow very lush lawn in our northern climates," Goggin observes. "It's scraggly, it doesn't have much to it, and so it's not even providing the function that a healthy grass would in absorbing water." In fact, if compacted from people walking and playing on it, and from baking in the sun, a lawn may offer little more runoff resistance than a driveway; the grass blades just lie down flat as water rolls over them, especially if coming from a concentrated source like a home's downspout.

Beyond the impacts of runoff, the loss of natural lakeshore habitat and the hardening of shorelines does wide-ranging, lasting, and cumulative harm to the ecosystems of lakes and their surroundings. Julia Kirkwood, who works in the Nonpoint Source Program with the Michigan Department of Environment, Great Lakes, and Energy, points to research in Michigan, Wisconsin, and Minnesota that demonstrates the effects. She mentions sea-walls, rock riprap, and removal of vegetation from the water and the near-shore land as pervasive on some lakes.

"We were seeing huge impacts," says Kirkwood, who also chairs the Michigan Natural Shoreline Partnership, a collaborative group that promotes ecologically sound shoreline property practices. "Some of our lakes are about 90 percent shoreline hardened. We have a lot of properties where people think they need a seawall, but they really don't." One aim of the partnership is to promote the use of vegetation or natural materials in place of rock and steel to prevent shoreline erosion, and to create transition zones of natural habitat where water meets land. In addition, the partnership launched the Michigan Shoreland Stewards program to help individual lake property owners assess their properties and management practices.

Fish and Wildlife Impacts

A paper from the University of Wisconsin–Stevens Point Center for Land Use Education puts the effects of shoreline hardening in clear perspective:

"The quality of fish and wildlife habitat generally decreases as the density of development increases along shorelines. Changes in water quality, bottom sediment, water levels and terrestrial and aquatic vegetation all contribute to this decline."[2]

Natural lakeshores are rich in habitat: the tree canopy providing nest sites for songbirds; the understory enabling travel corridors for mammals like foxes, minks, and fishers; dead trees and snags creating perches and nest sites for owls, wood ducks, and woodpeckers; the ground cover hosting thrush and ovenbird nests; shoreline grasses offering shelter and forage for small mammals like shrews, weasels, and mice. These values are compromised when shoreline habitat is lost or fragmented. The Center for Land Use Education paper cites various scientific studies that document the impacts. For example:

- Many lake property owners remove aquatic plants to make way for swimming areas, piers, and boats, removing crucial habitat for fish, waterfowl, and shorebirds. A study on lakes in northern Wisconsin found that developed shorelines with an average of one home for every 330 feet of shoreline had 92 percent less coverage of floating-leaf plants and 83 percent less coverage in emergent plants than shorelines without development.[3]
- Another northern Wisconsin study found that "city birds" such as crows and goldfinches became more common along developed shorelines, while insect- and fish-eating species such as warblers and loons declined.[4]
- A study on three Minnesota lakes found that black crappies and smallmouth bass were far less likely to build spawning nests along developed shorelines than natural shorelines—only 74 of 852 crappie nests were near shorelines with any kind of development.[5] More broadly, a study correlated increasing impervious surfaces in a watershed with fewer fish species in lakes and fewer fish overall.
- Green frogs declined when shorelines were developed; the frogs disappeared from shorelines where lots had only one hundred feet of lake frontage, likely because of lost habitat and conditions that made them more vulnerable to predators.[6] "Green frogs also vanished from sites when housing density met or exceeded about thirty-three houses per linear mile of shoreline," says Goggin. "Our zoning rules in Wisconsin allow for about fifty-two homes per linear mile and so are not adequate to abate the cumulative impact of unsound building patterns, which don't account for runoff and the conservation of natural vegetation."

Studies found that sediment on walleye spawning areas coming from impervious surfaces and other sources can cause the eggs to die from inadequate water circulation and too little oxygen, putting healthy walleye populations at risk.[7]

The loss of fallen trees, branches, and other woody material from nearshore water also harms fish and wildlife habitat. Lakefront homeowners often clear wood away to improve swimming areas or make way for piers. If a tree falls into the water, the impulse is to cut and remove it. As a result, many developed shorelines are nearly devoid of fallen wood. A study by the Wisconsin Department of Natural Resources shows the consequences. The lake studied was naturally rich in woody habitat, with about 750 logs per mile of shoreline. From one of the lake's two lobes, the researchers removed about 75 percent of the wood, reducing it to a level consistent with lakes that have modest numbers of homes and cottages. After four years, where the wood was removed, the yellow perch population collapsed; the perch lost spawning habitat, and the young that hatched were far more vulnerable to predators. Largemouth bass, with few perch to feed on, began to eat their own young and eat more frogs; their growth rate declined.[8]

Scenic quality, a major attraction to northern lakes, can also suffer when lake properties are overdeveloped. In a Minnesota survey, lake residents and visitors mentioned cabin and home development more than 85 percent of the time as a reason for loss of scenery. They also frequently mentioned docks and boat lifts and the removal of shoreline trees and shrubs.[9]

Seeking Remedies

While the degradation of lakeshores is pervasive, the remedies are in large part fairly simple, inexpensive, and widely known. The watchword is to think about the lake as a whole, to consider its entire human and natural community, to go lightly and develop gently. Arguably the simplest and most effective way to protect a lake is to maintain or plant a shoreline buffer— a strip of natural vegetation extending twenty feet or more up from the water. If the lakeshore is already natural and intact, all that's necessary is to let it alone and be watchful for invasive species like reed canary grass or purple loosestrife.

If there is an existing lawn, as a first step Goggin recommends no-mow: "Just put that lawn mower away and see what's in the native seed bank. Often there is enough dormant native seed still in the lakeshore area so that the plants will return and start to restore some of that natural habitat."

A developed but largely natural shoreline helps protect the lake from runoff and nutrient pollution. (Ted Rulseh)

A next step is some form of accelerated recovery, establishing a suite of native plants, from ground cover, to shrubs, to trees.

While none of this is complicated, the challenge lies in encouraging lake property owners to change entrenched attitudes and behaviors and adopt lake-friendly practices. For example, Goggin observes that lawns are deeply ingrained in the American psyche: "If you were a post–World War II person or family, one way you showed the world you had made it was by having a piece of green lawn. But we can't bring that suburban ethic and that lawn mentality to the lake edge, because it means dirty water. Eventually it means algal blooms. It means your kids can't swim in the lake, you can't boat in the lake, you can't enjoy the lake the way you used to."

Paul Radomski, a research scientist with the Minnesota Department of Natural Resources and the state's leading expert on lakeshore habitat management, says the Kentucky bluegrass used for many lawns "has done more to damage lakes than any other plant." Radomski sees more lake residents today restoring their shorelines or leaving substantial portions natural, although there are geographic differences: "Closer to the Twin Cities, we see a higher occurrence of lawn-to-the-lake with no regard to trees or shrubs for screening. But in some counties in the lake region of this state, there are high rates of natural shorelines."

Grants and More

To help bolster natural shorelines, state agencies offer a variety of education, research, and grant programs. For example, the Michigan Natural Shoreline Partnership encourages sustainable shoreland practices. It includes the state Department of Environment, Great Lakes, and Energy; Michigan State University Extension; Michigan Conservation Districts; the Michigan Lakes and Streams Association; and other educational and advocacy groups. It looks to train contractors and landscapers about shoreline technologies and erosion controls, educate property owners about natural shorelines, and support government policies that promote sound and sustainable shoreline management.

In Wisconsin, the Department of Natural Resources' Healthy Lakes and Rivers program gives grants of up to $1,000 to property owners for the following improvements:

- Fish sticks (cluster of trees placed in the water to create fish habitat)
- Native plantings of trees, shrubs, and other vegetation alongshore
- Upland rain gardens to capture and absorb runoff before it reaches the lake
- Diversions to direct water to where it can safely soak into the ground
- Rock infiltration pits to trap and filter runoff from roofs, driveways, and other surfaces.

From 2015 through 2020, grants had been issued in 29 counties, on 536 properties, on 66 lakes, accounting for 978 best practices (212 fish sticks, 96 diversions, 82 rock infiltrations, 423 native plantings, and 165 rain gardens).

Minnesota Department of Natural Resources offerings include a Restore Your Shore tool for property owners and service professionals to use in restoration and protection projects. Designed to help build deeper understanding of lakeshore ecosystems and natural shoreland management, it includes case studies on innovative restoration projects, a Native Plant Encyclopedia from which landowners can choose plants well suited to their locations, and a step-by-step guide for designing and completing projects.

Numerous counties, lake organizations, watershed councils, and other groups also offer a host of resources: educational brochures, workshops, websites, videos, and even monetary rewards for natural shoreland projects. For example, the Burnett County Shoreline Incentive in northwest Wisconsin pays for 70 percent of the initial cost of plants and materials for projects that help restore or preserve natural waterfronts. Participants receive a free site visit by a natural landscape expert, a planting plan, a special shirt or cap, and a sign to display at the water's edge. More than 750 properties have been enrolled since the program began in 2000.

Promoting Best Practices

While accomplishments like these are encouraging, they pale beside the persistent threats to the tens of thousands of lakes in the three states. Radomski advocates for persistence in delivering the message of shoreline protection and for a slow, steady process of breaking down the barriers to improvement. "A lot of lake advocates and lake association leaders have done wonderful things in communicating to their memberships that if we really care about water quality and about the lakes, then here is why we need to think about restoration," Radomski says. His state of Minnesota has reached out to landscape companies, encouraging native plantings and promoting alternatives to rock riprap as protection against shoreline erosion.

"Our watershed districts and soil and water conservation districts hold training workshops with contractors," Radomski explains. "Together we educate them about restoration of shorelines, how to do it, what kinds of designs, what kinds of plants, where to get the plants. A lot of people want the work done for them; they don't want to do it themselves. When landscape service providers see a growing contingent of lakeshore owners interested in more sustainable designs, that lowers a barrier. What's the next hurdle? People can't find any native plants in a nursery? Well, we've worked with the nurseries. As you identify the barriers, then one by one you can start addressing them."

Radomski concludes, "We are inseparable from the rest of nature. If nature degrades around us, then our lives will be worse for it. Our conduct is just if it does not harm the ecological health and beauty of a place."

CHAPTER 7

〜〜〜〜〜〜〜〜〜〜〜〜〜〜〜〜

Zoning and
Its Discontents

"I was stunned."

That's how John Richter describes his reaction in the summer of 2015 when an act of the Wisconsin legislature took away the authority of counties to enact shoreland zoning rules more protective of lakes than the state's minimum standards.

From the moment Governor Scott Walker signed the legislation, the same basic protections applied to nearly every lake in Wisconsin, no matter where located, no matter how large, how small, and how deep, no matter how environmentally fragile.

Richter saw the legislation as greatly weakening what had been an effective shoreland zoning program in northern Wisconsin's lake-rich Vilas County, where he owns a home on 1,057-acre Plum Lake. The legislation negated the county's lake classification system requiring different minimum lot frontages depending on lake size and sensitivity to development:

- Class 1: 150 feet minimum frontage, 20,000-square-foot minimum lot size.
- Class 2: 200 feet minimum frontage, 40,000-square-foot minimum lot size.
- Class 3: 300 feet minimum frontage, 60,000-square-foot minimum lot size.

It also negated the state's long-standing set of minimum standards related to lot size, building setback from the water, vegetation along the waterfront, filling and grading, and impervious surfaces. Suddenly, on Plum and the rest of the county's more than thirteen hundred lakes, the zoning department could not require lake lots with more than the state minimum one hundred feet of frontage. The county lost the authority to deny permits for most boathouses and could not require homes to be set back more than the state minimum seventy-five feet from the water. The new rules also weakened the county's ability to require vegetative buffer strips to impede the runoff of nutrients and other pollutants into the lakes. In sum, Vilas and seventy-one other counties lost much of their authority to enact ordinances they saw as needed to preserve lake water quality and lake ecosystems.

Richter was shocked not just by what the majority of legislators pushed through but by the way they did it: "in the eleventh hour and about the fiftieth minute," with no opportunity for public debate and with no separate public hearings or discussion in the state assembly or senate. Under Wisconsin's biennial budget process, legislators can insert nonfiscal items into the budget bill, which then receives an up or down vote in both legislative chambers.

State Senator Tom Tiffany and State Representative Adam Jarchow were the most responsible for inserting the shoreland zoning changes as an amendment to the 2015–17 budget bill. They did so during the deliberations of the Joint Finance Committee, after statewide public hearings on the budget bill had already been held. The committee approved the amendment, as did the full legislature, with large Republican majorities in both houses. The governor signed the budget bill (officially named Act 55), and beginning on July 24, 2015, the state minimum shoreland zoning standards applied throughout Wisconsin.

Tiffany and Jarchow claimed their action would bring uniformity to lakeshore zoning and balance environmental concerns against property owners' rights. They also called it a reaction to what they saw as a history of overreach on shoreland regulations by the state Department of Natural Resources.[1] Environmental advocates, lake associations, and many county zoning officials saw it as an unjustified weakening of protections for lakes, especially in counties with many lakes and where high-quality waters are the drivers of tourism-based economies. Wisconsin Lakes, the statewide lake association, adamantly opposed the changes for abandoning the state's decades-old system of county-based shoreland and lake management.

For Richter, the change was personal. The Plum Lake Association, of which he was president, was concerned about excessive development, and the new regulations reduced the minimum lot frontage on the lake from 150 to 100 feet. "The state minimum became the state maximum," Richter says. "And it was not just about lake frontage. There were a number of other things that made it more difficult to protect the sensitive areas on the lakeshore. It took away local control. That was the primary rub. In Vilas County, where taxes on lakeshore property constituted 75 percent of the tax base, we didn't have control of the lakes anymore. Over time, it will have a negative impact on the quality of the water. I was stunned that such a stupid thing could have happened."

In response, Richter set out to seek a rollback of the new provisions. He and a handful of supporters solicited funds from lake associations, other groups, and some local government bodies and formed the Wisconsin Shoreland Initiative, which then hired a lobbyist and an attorney in Madison, the state capital. Despite their efforts, a bill to repeal the Act 55 provisions never made it to the senate or assembly floor.

In the late 1960s, as the Baby Boomers matured and bought up lake properties, development threatened to overwhelm the lakes in the North Country and elsewhere. In response, state and local governments looked to apply a brake, and the tool they chose was shoreland zoning—special rules to govern subdivisions and other building on land on or near lakes, rivers, and streams.

The Wisconsin experience demonstrates that special zoning has only limited power to ensure ecologically sound lakeshore development. Minnesota and Michigan have their own approaches to shoreland regulations; natural resource officials from both states admit that there are gaps in protection. In particular, zoning has been marginally effective in preserving natural vegetation in near-shore areas.

Across the three states, shoreland zoning administration is left to county and other local governments. A key challenge in several counties is the sheer number of lakes and lakefront properties. For perspective:

• Michigan's Oakland County has more than four hundred lakes, Iron County more than three hundred, Gogebic and Schoolcraft about two hundred each. The northern Lower Peninsula's Tip of the Mitt Watershed Council has some two hundred lakes of varied sizes in its four-county service area.

- Wisconsin's Oneida County has more than eleven hundred lakes, Vilas County more than thirteen hundred, Sawyer County some seven hundred.
- Minnesota's Itasca County has more than fourteen hundred lakes, Otter Tail County nearly one thousand fifty, St. Louis County more than five hundred, Crow Wing County more than four hundred.

Some of these counties, and others, have thousands or tens of thousands of lakefront properties. A related challenge is the rate at which those properties are being purchased and developed and existing homes remodeled and expanded. Many county zoning departments lack the staff to thoroughly inspect these properties for compliance with building permit provisions. For example, the Planning and Zoning Department in Oneida County, Wisconsin, has just four zoning technicians and two land use specialists to cover not only shoreland properties but all subdivisions and building on land under the county's jurisdiction. Meanwhile, many property owners alter their lakescapes—adding structures such as gazebos, patios, and stairways; cutting trees; and stripping near-shore vegetation—without applying for permits, or in the absence of requirements to do so.

Shoreland Zoning in Minnesota

In Minnesota, shoreland protection at the state level began in the late 1960s. "The lake country was being heavily developed and its character was changing quickly," notes Daniel Petrik, a land use specialist in shoreland and river-related programs for the state Department of Natural Resources. "People didn't like what they were seeing; the unplanned growth was a cause for concern," he says. After a bipartisan push supported by corporations, small businesses, home builders, environmental groups, local governments, and others, the state created its first shoreland management rules for counties in 1969. Rules for cities were enacted in 1976, and in 1989 the rules were updated and unified to apply to both counties and cities.

"The approach today is one of state government partnering with cities and counties to communicate best management and regulatory practices that address local challenges," Petrik observes. The 1989 rules are dated and no longer address many current development issues. For example, according to a DNR source, in 1989 there was no broad recognition of climate change; homes built in the lake country were still somewhat modest in size, and the style of landscaping was fairly basic. Today Minnesota has changing precipitation regimes, warming lakes, and changing ecosystems and species mix, some species in retreat and some in advance. Homes are getting larger; many people are landscaping their lake properties in a more suburban style and using the near-shore area more intensively. The state's shoreland regulations were not developed to address those changing impacts, and the DNR lacks the authority to update the rules unless so directed by the state legislature.

The statewide rules govern shoreland development under a three-part lake classification system established in the late 1960s. The smallest lot sizes, lake frontages, and structure setbacks from the water are allowed on lakes classed as general development—those already developed, with significant upland areas around them, and considered large and deep enough to be relatively resilient to development. The greatest lot sizes, frontages, and setbacks apply to lakes classed as natural environment; they are surrounded by generally flat terrain, are relatively shallow, and have little or no development. In between are lakes classed as recreational development—lakes of medium size and depth and with modest upland surroundings.

Petrik notes that the state's dimensional standards such as lot sizes and setbacks are reliably enforced in the local building permit process. However, enforcement is difficult on what he calls the most important aspect of lake protection: preserving natural vegetation in the near-shore area. "With thousands of miles of shoreline in some counties, it is very difficult for local zoning staff to monitor and enforce limitations on vegetation cutting after a home is built," Petrik says. The rules prohibit intensive vegetation clearing within the near-shore area and on steep slopes but allow limited cutting and trimming for stairways and recreation areas such as beaches, and to provide a view of the water from the home.

"Many Minnesota counties are very large areas," Petrik says. "Counties even in good times do not have enough staff and must focus their enforcement on a responding-to-complaints basis." Some Minnesota counties have taken an innovative approach to shoreland regulations by putting in place a riparian vegetation standard. Under this approach, restoration of shoreland becomes an enforceable condition of a building permit. Itasca County, for example, requires a shoreline buffer that ranges from ten to fifty feet up from the water, depending on the lake classification. Petrik states, "If you pull a building permit in Itasca County, and you're on a lake that requires a ten-foot buffer and you don't have one, you will need to restore a ten-foot buffer as a permit condition."

Itasca's rules illustrate that in Minnesota, local governments can enact standards more protective than the state's, subject to DNR approval. Petrik observes, "Given that the state rules are more than thirty years old, many counties and some cities are saying, 'We need to change our own rules to address evolving challenges and the lake protection issues we're seeing.' They can take action to protect their own natural and economic resources."

An Innovative Shoreland Standards Showcase on the state's shoreland management website recognizes counties that have adopted more protective standards. For example:

- Crow Wing County limits patios in the near-shore zone to 250 square feet. Otherwise, impervious surfaces are not allowed within that zone except stairways, lifts, or landings.
- In Aitkin County, rock riprap on shorelines is allowed only where established erosion problems cannot be corrected through more natural methods. Installation of riprap requires a vegetative buffer to a distance from the water determined by the county staff.
- Beltrami and Washington counties and the City of Detroit Lakes require home setbacks from the water larger than the state shoreland standards.

Shoreland Zoning in Michigan

Michigan's shoreland protection program gives localities more flexibility than does Minnesota's. The Michigan Department of Environment, Great Lakes, and Energy directly regulates and permits activities in wetlands close to or connected to lakes and streams, and activities such as dredging, filling, and vegetation removal below the lakes' ordinary high-water mark. However, it leaves regulation of near-shore activities on land to localities.

Eric Calabro, an inland lakes analyst with the department, observes that state statutes leave "some gaps for local governments to fill" in areas such as ordinances that govern stormwater management, nuisance plant control, and septic systems. The state sets no minimum setbacks, minimum lake frontages, or other specific requirements for development near shore; permits for such work are required but are administered locally. "There is a lot of room where the state doesn't regulate for localities to enact whatever they want," Calabro says. "That is valuable because Michigan has such a variety of lakes and a variety of uses. Local ordinances can be helpful in sculpting the community and local ethics as desired in those areas. A state permit is required for any work in wetlands and work at or below the ordinary high-water mark of inland lakes and streams. There may be local permit requirements in addition to what the state requires."

The state gives extensive guidance to localities in structuring shoreland zoning provisions. A lake protection guidebook for localities states, "Proactive efforts by local governments to preserve the quality of life in their communities are part of the rich history of home rule in Michigan. Beginning in 1921 with the City and Village Zoning Act, local governments in Michigan have had the authority to implement local regulations that foster the health and well-being of their communities. This includes conserving natural resources."[2] The guidebook describes state planning tools available to help communities develop ordinances that regulate wetlands, lake setbacks and lot widths, piers, soil erosion and sediment control, landscaping and impervious surfaces, septic systems, and more.

Aside from regulation, education and outreach are central to the department's mission. "The state is involved with multiple partnerships that promote healthy shorelines," Calabro says. These include the Michigan Natural Shoreline Partnership and the Michigan Inland Lakes Partnership, both organizations of diverse groups interested in protecting and preserving lakes. One popular offering is a certification course that teaches landscape contractors how to create natural shorelines and natural shore erosion protection for properties. "We have people come in from Michigan State University Extension," Calabro says. "State staff members present the regulations. University professors talk about plants and soils. Once the contractors complete the course and pass an exam, there is a field day where we typically do a shoreline restoration." The contractors are then designated as Certified Natural Shoreline Professionals.

Meanwhile, the adequacy of local shoreland protection varies greatly. For the benefit of county and local governments, the Tip of the Mitt Watershed Council in 2011 published gap analyses (strengths and weaknesses) of zoning provisions in its four-county territory in the northern part of the Lower Peninsula. Cheboygan and Emmet counties have zoning ordinances that include shoreland protections, as do some of their municipalities. Antrim and Charlevoix counties rely solely on municipal and township zoning regulations. The council noted that these two counties actively support local zoning initiatives but also observed: "For water protection efforts, in particular, there are situations where county zoning, or specific ordinances, offer more effective and coordinated protections. For example, townships may have shorelines on a common water body, but they can be uncoordinated in their protection efforts. One township could require protections that are strong or adequate, while the neighboring township does not require the same protective steps, at all. This does not help the water body in question."[3] Here are a few examples of local and county ordinance provisions that at the time the council considered effective:

- Evangeline Township (Charlevoix County) has a Natural Vegetation Waterfront Buffer Strip provision by which owners of properties being developed or redeveloped must keep or create a buffer strip at least twenty-five feet deep (fifty feet for lots with steep slopes) across the full width of the waterfront. This strip must be planted with native trees and woody shrubs to provide complete ground coverage except for an access path and a viewing corridor (a break in the shoreline trees and vegetation to provide a view of the water from the home).[4]
- Forest Home Township (Antrim County) addresses impervious surfaces by limiting private roads to eighteen feet wide in environmentally sensitive areas and requiring all private roads to have drainage plans. Parking lots must be landscaped, graded, and drained so that runoff is retained within the area.[5]
- Cheboygan County defines a Lake and Stream Protection District that includes all property within five hundred feet of the shoreline of any river, stream, pond, or lake. It requires all buildings, parking lots, and other impervious surfaces to be set back at least forty feet from the water (exceptions for boat docks and other water-related structures). The ordinance encourages the zoning administrator each year to publicize through local media the need for natural shoreline vegetation strips.[6]

In general, the council analyses characterized county and local regulations of impervious surfaces on lakefront properties as weak, where they existed at all.

Shoreland Zoning in Wisconsin

Wisconsin's experience with statewide, one-size-fits-all shoreland regulations has been mixed, at best, according to zoning administrators and lake advocates. "The change happened because of some very intense personalities in the legislature," observes Michael Engleson, executive director of Wisconsin Lakes. "I don't think the changes to the statute would have

Shoreland zoning in some localities allows clear-cutting of a corridor to create a view of the lake from the home. (Ted Rulseh)

occurred if not for those specific people. Shoreland zoning had been widely popular on a bipartisan basis, and especially the local control aspect of it. The issue became caught up in the broader politics of that period. It was not as if there appeared to be intense and widespread support among the Republican majority to make those changes."

Bob Martini, retired statewide river protection coordinator for the Wisconsin DNR and president of the Oneida County Lakes and Rivers Association, observes that the shoreland zoning rule change was just one of several adverse events occurring at the same time. "A whole series of problems hit at once," he recalls. "They imposed the statewide standards and severely eroded local control. At the same time the climate was changing. We have a lot more rain and much higher temperatures, and water temperatures are up. There's a longer growing season because ice coverage is shorter now than it has ever been. That means we have more favorable conditions for algae and plant growth than we had forty years ago."

He notes that the state legislature and governor cut the DNR staff, reducing regulatory oversight, technical assistance, and information around water resources. Citizen lawsuits were eliminated so that citizens lost the ability to sue the DNR in court for failure to properly implement the federal Clean Water Act. In turn, the DNR lost its authority to appeal county zoning it considered inadequate to protect water quality. Meanwhile, larger and more powerful boats proliferated at a time when lake water levels were approaching all-time highs.

The practical effects of the zoning rule changes have been substantial. Counties rich in lakes that each year attract hundreds of thousands of visitors have no more power to regulate shoreline development than do counties with, as one Oneida County Board Supervisor has put it, "two mudholes and a slough."

Jay Kozlowski, zoning and conservation administrator for Sawyer County, in the state's northwestern lake country and home to the tourist-hub city of Hayward, says the state legislators "should have done a little more scientific research on the effects of some of the legislation they passed. A lot of zoning staff would have liked to be at the discussion table when the legislature was trying to implement some of these changes, as ultimately we are the ones charged with regulation."

For example, counties now have only limited power to refuse permits for boathouses. "We've seen quite a few new boathouses," Kozlowski says. "Those are going to have a larger impact on lake ecology and the overall healthiness of the lakes, because the structures are located so close to the water, four or five feet from the ordinary high-water mark." Counties still can control the size of boathouses and can deny them in some limited cases, such as on sites with steep slopes or in floodplains. Otherwise, zoning agencies are required to permit them.

Natural shoreline buffers also suffered under the new regulations. "We were a little bit unique in Sawyer County in that any land use permit on shoreland zoning, even for a garage two hundred feet away from the water, would have required shoreline vegetation mitigation," Kozlowski says. "We wanted a buffer zone on all shoreline properties. That didn't always sit well with people, but we saw it as an improvement for keeping the natural scenic beauty of the area. That was stripped away." Now, except in special cases, property owners can choose from a checklist of options for mitigation that include a buffer zone but also adding a rain garden, removing an old steel septic tank, removing shoreline lighting, increasing the shoreline setback, and removing a structure within the shoreline setback zone. Furthermore, a property owner with an existing house or cabin much closer to the water than the seventy-five-foot setback can tear it down and rebuild in

In Wisconsin, local zoning agencies must permit boathouses except under a few special conditions. (Ted Rulseh)

the same footprint without doing any mitigation at all, Kozlowski says. And that new structure can be built as tall as thirty-five feet, creating some unusual configurations.

Kozlowski observes, "When I was reading through the draft language, I said, 'You're going to have someone tear down that old cabin that was right tight to the water. They're going to build the thing taller than it is wide, and it's going to tower over the lake.'" That very thing has happened: "There are two of them in particular. They tore down the cabin and built what now looks like a silo, thirty-five feet in height, ten feet off the water. They look quite bizarre to me. You can talk about the ecological effects, the runoff effects, the pollution effects, but also just the aesthetics. One family had a single-level cabin they used for a vacation getaway. Now they have the kitchen downstairs, a couple of bedrooms upstairs, and then a loft above that. It's a three-story structure that's crammed into a small footprint."

Kozlowski believes some changes under Act 55 are for the better. For example, property owners now can do small projects—up to two hundred square feet—on existing homes and cabins built years ago inside the seventy-five-foot lakeshore setback without having to go through variance procedures. "I kind of like the two-hundred-square-foot expansion," Kozlowski says. "A lot of people have been able to use that, and it has helped the economy up here. Many people who are doing those expansions are still doing buffer zone mitigation. That's helping the lake, and the property owners are getting the small addition they always wanted."

For the future, he's concerned that the new regulations negated the state's lake classification system with its larger minimum frontages and lot sizes on smaller and more environmentally sensitive lakes. "When this first came about, I thought we would see quite a bit of lot development and lot splits on those bodies of water," he says. "I thought we would start seeing the smaller lakes being overdeveloped, and we really haven't seen that trend yet. But fast-forward fifteen years from now and that could certainly be the case."

In Vilas County, in the north central area and Wisconsin's richest county in lakes, zoning administrator Dawn Schmidt shares the concern about excess development on smaller lakes. Immediately after Act 55, a few subdivisions with one-hundred-foot lots came in for review. In 2020, more such proposals came in, including some for lakes that previously required three hundred feet of frontage. "I have a great concern for the water quality," Schmidt says.

One reason is that smaller lots mean more septic systems and their pollution potential. "We're an area of numerous in-ground septic systems," says Schmidt. "We do not have a lot of municipal sewer. I tried to get across when they did Act 55 that they were going to overtax the ground with pollutants. Eventually that's going to affect our groundwater, and it will affect our lakes." Adding to the risk, the state legislature liberalized restrictions on short-term vacation rentals, often occupied by numerous guests and stressing the septic systems far more than when the homes or cabins were used only occasionally by one family.

As in Sawyer County, shoreland mitigation in Vilas County has suffered. Previously, any property owner building an addition was required to do mitigation, which often included a lakeshore buffer strip. "They could come to us with a plan and we could say yea or nay," Schmidt says. "Under Act 55, if they buy a home or cabin with a clear-cut yard, we cannot force them to replant it. If they build a new house seventy-five feet back from the lake, we cannot force them to mitigate." Mitigation is required only when a property owner adds to a home within the seventy-five-foot setback.

Schmidt also dislikes the provision that allows owners to "wreck and rebuild" structures within thirty-five feet of the shore, where their impact on water quality is the greatest. Previously, such structures were gradually going away because only minor expansions and upgrades were allowed; owners who wanted larger homes most often had to build them at least seventy-five feet from the water. "So now, those structures are virtually never going to stop being there."

Issues with Enforcement

Above all, Schmidt regrets the erosion of the local control county zoning departments once enjoyed. "There are so many different areas in Wisconsin that we need to be able to protect locally," she says.

Martini points to two other problems with the statewide zoning rules. Limits on impervious surfaces designed to restrict runoff from properties into lakes were weakened, and so were rules related to viewing corridors. "Previously, you were allowed a thirty-foot-wide viewing corridor, but you still had to maintain a thirty-five-foot buffer zone up from the water's edge," Martini says. "Now you can cut out the buffer within the viewing corridor, and you can have thirty-five feet of viewing corridor for every hundred feet of frontage." That means someone who owns three hundred feet of frontage can clear-cut a viewing corridor 105 feet wide. "And typically, once the trees are cut, they want to change the vegetation to grass or something exotic instead of natural vegetation," Martini says.

Compounding all these concerns, most county zoning departments lack the resources to enforce aggressively the rules they still have. Kozlowski observes, "We're basically limited to complaint-driven enforcement. If we see blatant violations, we'll do enforcement on those, but we don't have the time or manpower to drive down every little back road or to hop in a boat and cruise along the shorelines of the lakes to look for violations. We are grossly understaffed to do that kind of enforcement."

The net effect, Martini says, is that

> instead of responding to the public's consistent desire for better water quality, every change in the past ten years has been allowing for less water quality, on the argument that with more development the economy will be better, and there will be more jobs. Now we've got shoreland zoning that doesn't really protect anything. The economy in northern Wisconsin depends on good-quality lakes, and yet the legislature did just the opposite of what you would normally do when trying to protect an economic asset. They allowed less protection for what is arguably the most important economic asset in northern Wisconsin. It defied logic, and it defied science.

Kozlowski echoes the concern.

> Zoning administrators who have been in the game longer than I've been around have seen the pendulum swing. Shoreline zoning became very restrictive in the 1990s and 2000s, and since Act 55 the pendulum has gone back to having more flexibility for shoreland regulations. Will that pendulum swing back again someday? It really depends on whether there is a large grassroots movement that persuades legislators to start having some stricter enforcement. We have a lot of recreational-type seasonal tourists who come to the area, and that's because of our lakes. If our lakes start to degrade, there goes our tourism and revenue for the county. I just hope people will keep that in mind for another fifty to a hundred years down the road.

CHAPTER 8

Defective Septic Systems

How Big a Problem?

In the late 1970s, water quality was steadily declining in the lakes around the City of Otter Tail in west central Minnesota, and property owners were concerned. A group began looking into the cause and asked Otter Tail County Land and Resource Management to investigate.

High on the list of suspects, along with less-than-optimal farming and shoreland development practices, were the septic systems around the 13,700-acre Otter Tail Lake and five smaller lakes nearby. State and federal grants helped fund inspections of the old septic systems. Otter Tail County staff and private contractors then completed a study suggesting that phosphorus from those septic systems was likely a major cause of the lakes' decline.

An environmental impact statement completed later raised an added concern about high levels of fecal coliform bacteria—also coming from the septic systems—in the lakes' near-shore water. Based on an engineering firm's estimate, the cost to replace the septic systems with sanitary sewer was prohibitive. In response, the county commissioners formed the Otter Tail Water Management District to devise and implement a cost-effective remedy.

"The water quality had gone downhill," recalls Rollie Mann, administrator of the district from 1984 until he retired in 2019. "During the wintertime, there had been in-depth studies along the shoreline where they drilled holes and checked for water quality. They were finding discharges from septic systems, many of which were basically fifty-five-gallon drums sunk into the ground. The old adage was, 'Dig until you hit water, put your fifty-five-gallon drum in, and you'll never have to pump it.' There were also a few straight pipes discharging directly to the lakes. It turned from a pollution issue into a health issue in that high fecal counts were being found in the swimming areas around the lake."

The Otter Tail district story provides lessons in both the effects substandard septic systems can have on lakes and the benefits that creative, responsible septic system management programs can deliver.

~~~~~

Imagine, around a lake, a dozen or more continuous streams of sewage-contaminated water flowing into areas where people wade and swim and where anglers cast for bluegills or bass. This scenario may be overly dramatic, but it serves to illustrate the potential effects of defective septic systems around North Country lakes. Except in the few places where municipal sewer systems have been installed, the homes and cabins around lakes in Minnesota, Michigan, and Wisconsin rely on some form of on-lot system to collect and treat toilet flushings, bath and shower water, dishwater, laundry machine and water softener discharges, grindings from garbage disposals, and other wastes.

Some of these, and on some lakes the majority, are newer and properly functioning septic systems installed and maintained according to the latest codes. But others are septic systems installed thirty or more years ago when state and county regulations were less strict than they are today. While age does not necessarily correlate with poor performance, some older systems may no longer treat waste effectively; some likely have failed and are adding pollutants to the environment. Still others are old cesspools or seepage pits that never provided more than minimal treatment. Taken together, these systems can pose a significant threat to lake water quality.

Septic system maintenance professionals will attest that people from the cities who move to lake properties know little or nothing about septic systems and how to maintain them. Systems that fail through neglect or old age can leak pollutants into the groundwater and, ultimately, the lake. Worse, sewage from failed systems can emerge on the surface and be carried straight into the lake by runoff from rainfalls or snowmelt, adding nutrients, disease-causing bacteria (pathogens), and other contaminants to the water.

A study conducted by Michigan State University researchers found signs of pollution from septic systems to be more widespread than expected.[1] The study sampled sixty-four rivers that drain 84 percent of Michigan's Lower Peninsula, looking for E. coli and another common fecal bacteria. "All along, we have presumed that on-site wastewater disposal systems, such as septic tanks, were working," noted Joan Rose, Homer Nowlin Endowed Chair in Water Research at the university. "But in this study, sample after sample, bacterial concentrations were highest where there were higher numbers of septic systems in the watershed area. . . . This study has important implications on the understanding of relationships between land use, water quality and human health as we go forward. Better methods will improve management decisions for locating, constructing and maintaining on-site wastewater treatment systems."[2]

How They Treat Wastewater

To appreciate the importance of properly designed septic systems and responsible system care, it is useful to understand how these systems work. When properly designed, installed, and cared for, septic systems perform the same basic functions as a municipal wastewater treatment plant: removing contaminants and pathogens from sewage. The key difference is that while wastewater treatment plants employ sophisticated machinery and processes to treat large volumes rapidly, septic systems rely on the soil and the microorganisms living in it to remove substances that would otherwise cause water pollution.

A septic system is more than a septic tank. The tank plays an important role by capturing wastewater from inside the house, trapping grease and

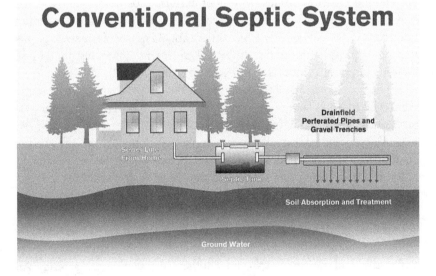

Conventional Septic System

Drainfield
Perforated Pipes and
Gravel Trenches

Sewer Line
From Home

Septic Tank

Soil Absorption and Treatment

Ground Water

(Eric Roell)

oils, and allowing the solid matter to settle to the bottom, where some rudimentary treatment occurs. But the real work of treatment gets done in the drainfield, which most often consists of a set of covered trenches in the soil, filled with rock usually 0.75 to 2 inches in diameter, or with some synthetic material that creates space for air circulation. As water from the house flows down a sewer pipe and enters the septic tank, the same amount of water (effluent) flows out of the tank and into the drainfield, where bacteria in the soil go to work, essentially eating up the impurities and also helping to remove pathogens.

The septic tank effluent contains organic matter that is dissolved in the water or floats in the water as tiny particles (suspended solids). Taken together, this material comprises biochemical oxygen demand (BOD); it is consumed by soil bacteria in the presence of oxygen supplied from the air spaces in the trenches. Oxygen must be able to move through the soil around the outside of the trench; if effluent becomes ponded in the trenches, bacterial action is curtailed (although oxygen still can reach the trench bottom by traveling along the sidewalls).

Septic systems in good working order do a fine job of catching and breaking down the organic matter and filtering out pathogens (see table 2). Significant treatment occurs in the biomat, a rich film of microorganisms that over time grows on the bottom and sides of the trenches. Still, the treated water that ultimately soaks through into the soil contains nitrogen and phosphorus. Once outside the drainfield trenches, these nutrients are headed down through the soil and in the direction of the water table.

For this reason, regulations, which differ by state and locality, specify a certain vertical separation (infiltration zone) that must exist between the

Table 2. Conventional septic system efficiency in removing pollutants

	Percent removed*
Biochemical oxygen demand (BOD)	>90%
Nitrogen	10–20%
Phosphorus	85–95%
Fecal coliform	>99.9%
Organic chemicals (solvents, pesticides, etc.)	>99%

* Assumes percolation and treatment in a 3- to 5-foot infiltration zone below the drainfield trenches)

Source: U.S. Environmental Protection Agency, 2002.

bottoms of the trenches and the seasonal high-water table, or a layer of impermeable soil, such as clay. This separation distance, typically three feet, helps protect the groundwater from pollution and so protects the lakes, toward which the groundwater gradually flows. In general, the greater the separation, the longer it takes for the nutrients to reach the groundwater.

Evaluating Effectiveness

The septic systems on most lake properties are major improvements over outmoded facilities like cesspools and seepage pits. Cesspools are cylindrical structures dug into the ground that simply receive raw wastewater, trap the solids, and allow the liquids to trickle into the soil. Seepage pits are similar except that they receive septic tank effluent from which the solids have been largely removed. A problem with both is that the wastewater is concentrated in a small area (perhaps a six-foot-diameter circle) and so can percolate rather quickly down toward groundwater. Septic system trenches, meanwhile, spread the tank effluent over a much larger area of soil—the number and length of trenches depending on the number of bedrooms in the home. This puts a much larger volume of soil to work treating the wastewater, while also helping to slow the downward movement of nutrients and other contaminants.

As table 2 shows, conventional septic systems are not effective in removing nitrogen, which is converted in the soil to nitrate that readily dissolves in water, enters the water table, and is carried toward the lake. The good news is that in most inland lakes, nitrogen additions are not a major cause of weed and algae growth (although high levels of nitrate in groundwater can pose risks to health). The even better news is that septic systems generally do a good job of trapping phosphorus, the main nutrient of concern in lakes. Phosphorus tends to attach itself (adsorb) to soil particles in and below the drainfield trenches. This takes 85 to 95 percent of the phosphorus out of circulation, and it takes a very long time for any of the phosphorus to reach the lake. However, there are exceptions. For one, phosphorus is more likely to find its way to the lake if the soil is coarse and sandy, and if there is a short distance between the septic system and the lake. More important, phosphorus becomes a threat if the septic system has been allowed to fail through old age or neglected maintenance.

There is no way to generalize how much phosphorus and other pollutants septic systems contribute to lakes, according to Sara Heger, an engineer, researcher, and instructor in the Onsite Sewage Treatment Program at the University of Minnesota Water Resources Center. She notes that phosphorus from septic systems needs to be considered in the context of all inputs to a lake: nearby cropland, shoreland runoff, fertilizer, and others. On a given lake, septic systems may or may not be important phosphorus sources. Much depends on how committed property owners and lake associations are to managing septic systems effectively. Age alone does not make a septic system a source of pollution. If installed in soils well suited for the purpose and if properly maintained, septic systems can treat effectively for a long time. A common rule-of-thumb life expectancy is twenty-five to thirty years, but many can last longer, especially if serving homes or cabins used seasonally. Heger observes, "Typically, seasonal homes have pretty low infrastructure. People go there to hang out, but they're not living there, and so the inputs to the system are less. A lot of seasonal places don't have a washing machine." Those systems could last forty years or longer.

Phosphorus Contributions

On the other hand, failed septic systems, along with cesspools and seepage pits, some of which still exist along lakeshores, can be continuous phosphorus contributors. "Phosphorus adsorbs well to the soil, but the soil has to be dry," Heger says. "If the soil is saturated, the phosphorus isn't coming in contact with the soil particles, and you're not going to get phosphorus removal." A failed system creates a plume of phosphorus slowly moving through the groundwater to the lake. The impact does not stop when that system is replaced. "Those plumes continue to travel, and the phosphorus that's in the water body continues to cycle, especially in lakes that don't have a stream outlet," Heger says. "Even in lakes that have an outlet, the vegetation holds on to the phosphorus until it dies in the winter, and then it's rereleased. So it can take years to decades for lakes that have had significant phosphorus inputs to clean up."

A key question is just how much phosphorus malfunctioning or failing septic systems can contribute to a lake. And as one might expect in these times, there's an app for that—actually a spreadsheet tool created by the University of Minnesota. Called the Septic System Improvement Estimator, it helps calculate how much phosphorus and other pollutants a septic system can release, depending on how well it is designed, installed, and maintained and is functioning.[3] Conversely, the tool can help predict how much less pollution a lake will receive if failing systems around it are repaired or replaced. The tool is designed mainly for professionals working in septic system management, but it does provide information that can help the average person understand septic system impacts.

Suppose, for example, that you live year-round in a three-bedroom lake home with a septic system that is being cared for properly, complies with all the regulations, and provides at least three feet of separation between the bottoms of the drainfield trenches and the groundwater. According to the spreadsheet, that system's annual phosphorus release is zero. Now suppose that very same system has only two feet of separation from groundwater. The tool says that system releases four pounds of phosphorus per year and also contributes *E. coli* and other bacteria of the types commonly found in human waste. Systems like this are insidious because they show no obvious symptoms: wastewater does not surface in the yard or back up in the house. The toilet still flushes, but partially treated sewage is headed for the groundwater. Going a step further, suppose that system is in such a condition that it no longer protects the groundwater at all. In that case, it releases four pounds of phosphorus per year, plus bacteria, plus amounts of BOD and suspended solids, much of which will eventually make its way to the lake.

Four pounds of phosphorus added to an entire lake may seem trivial, but its impacts are not. Recalling the Algae Adage (chapter 5), every pound of phosphorus that enters a lake can feed as much as five hundred pounds of algae. One poor-performing septic system releasing four pounds of phosphorus could help generate two thousand pounds—a ton—of algae in one year. From there it is easy to imagine the effects of five, ten, twenty, or fifty such systems around a lake. And surveys demonstrate that many lakes do have multiple failed or otherwise deficient septic systems on their shores.

An emerging concern is the recent trend toward wetter years, potentially elevating water tables and reducing the vertical separation between groundwater and septic system trenches. Heger observes, "We don't have enough data to say that water tables are actually rising. The general trend has been wetter, but to say for certain that's going to cause a rise in our water tables, we haven't had enough data or enough time. But if water tables are rising, that is not good for septic systems."

Jim Anderson, retired director of the University of Minnesota Water Resources Center, adds, "Lakeshore owners need to recognize that the water table affecting their septic systems is usually connected to the lake. So if their lake levels are rising, the systems that had adequate separation twenty years ago may not have it now."

Frequency of Failures

It is difficult to pinpoint what share of septic systems around the lakes are failing. On a large scale, the EPA estimates that one in five of the nation's homes (about twenty-five million) are served by on-site treatment systems rather than municipal sewers. Estimates of the prevalence of failing systems vary widely. The Michigan Department of Environment, Great Lakes, and Energy estimates that 30 percent of that state's homes and businesses have septic systems. "The failure rate for these systems in any given year is 5–10%, based on information from local county health departments submitted to the [Michigan Department of Environmental Quality]."[4] According to University of Michigan Extension, "Public health officials believe reported septic system failures represent only a fraction of the total number of failures statewide, and many go undetected or remain unreported for years."[5]

In reality, the only way to determine how many and which systems are failing in a locality is to inspect the systems individually. Here again, data is scarce and scattered. In a survey in Wisconsin's Door County, about one-third of 6,450 septic systems inspected by the county sanitarian staff were classified as failing to work.[6] (Door County is arguably a special case because the area is underlain by limestone bedrock covered in many places by relatively thin soil.) Michigan's Washtenaw County conducted an eighteen-month study after passing a county ordinance in 2000 that required drinking water wells and septic systems to be inspected at the time of the sale of a property. The study found that 18 percent of the septic systems inspected were failing or inadequate.[7]

One County's Experience

Among the more extensive septic system inspection programs is a survey initiative in northwest Wisconsin's Sawyer County. There, the county sanitarian's office has conducted on-site inspections of septic systems around sixteen lakes since 1991. "It has been an awesome way to educate, reeducate, and inform property owners who come up here to the rural world from urban settings, and all too often are not familiar with septic systems," says Eric Wellauer, county sanitarian. The surveys have led to replacement or repair of a number of systems that were failing.

Merton Maki, certified soil tester and Wellauer's predecessor as county sanitarian, says the impetus for the surveys came from lake associations whose leaders "saw a need to inspect some of these old, old systems that were put in during the 1940s, '50s, and '60s. They were interested in keeping

the water quality clean." Many of those old systems were cesspools or seepage pits, and often the property owner dug them down to the groundwater.
Inspections turned up many fifty-five-gallon barrels used as septic tanks;
these typically rusted out within eight to ten years after being installed.
Some conventional septic systems with drainfields were also found to be
failing, the trenches full of water and in some cases, sewage flowing on
the surface. "Our highest percentage of failures around a lake was probably
30 to 35 percent," Maki says. "Those were normally on smaller lakes in rural
areas where, back in the old days, a lot of things were done without permits. It's a very good program, and to my knowledge Sawyer County was
the only county that was doing a program like this."

To qualify for a septic system survey, a lake association must get consent
from a majority of property owners. Then, in theory, all property owners
on the lake must allow an inspection, although a few still refuse permission. A thoroughly trained summer intern, supervised by the county sanitarian, inspects the systems. The process includes an interview with the
property owner and an inspection of the drainfield. If necessary, a soil bor-
ing is performed to check whether the required three feet of separation
exists between the bottom of the drainfield and a limiting factor: groundwater, bedrock, or saturated soil. If not, the system is recorded as failed. In
addition, the sanitarian's office relies on state law, which describes failure
as the presence of any of five conditions:

- Discharge of sewage into surface water or groundwater.
- Introduction of sewage into zones of soil saturation that adversely affect
  system operation.
- Discharge of sewage to a drain tile or into zones of bedrock.
- Discharge of sewage to the surface of the ground.
- Failure to accept sewage discharges and backup of sewage into the home.[8]

Table 3 shows the results of the four most recent septic system surveys.
In total, 70.1 percent of inspected systems passed, while 14.6 percent of
systems either failed or had inconclusive results. Some of the latter likely
are defective or failing, Wellauer says.

"In an ideal situation, after an inspection on a given property, we'd like
to have a pass or a fail," Wellauer says. "Unfortunately the world we live in
is not that simple." For example, the surveys did not look in depth at the
tanks: "We relied on licensed individuals who pump and maintain the tanks
to let homeowners know about the condition of their septic tanks, pump

Table 3. Septic survey results, Sawyer County, Wisconsin

Lake	Pass	Fail	Inc	OFC	Refused	Vacant	Total Properties
Nelson (2010)	338	44	31	6	36	N/A	453
Barber (2011)	72	6	2	5	6	N/A	91
Big Sissabagama (2017)	132	15	28	3	N/A	41	209
Lac Courte Oreilles (2013)	558	38	65	106	12	124	815*
TOTAL	1,100 (70.1%)	103 (6.6%)	126 (8.0%)	120 (7.6%)	54 (3.4%)	165 (10.5%)	1,568

Inc = Inconclusive

OFC = Order for correction

* Total Properties Inspected figure is less than the total of the category results because the 106 orders for correction included some of the systems in the other categories.

Source: Sawyer County (Wisconsin) Zoning and Conservation Department.

tanks, or holding tanks." Old steel septic tanks, which are prone to rust-through, leakage, and collapse, were automatically designated as failed.

Beyond that, the inspections focused on the drainfields. In some cases, changes to landscapes, damaged or missing components, and other factors made it impossible to locate or properly inspect the field and state with certainty whether, for example, the needed separation from groundwater existed. "We did send out letters informing the owners of the result and telling them that they should have some care and concern and find out more about their systems," says Wellauer. "Some did and some did not. For my own purposes, I divided those systems into two categories: inconclusive pass and inconclusive fail." Owners of failed systems were ordered to repair or replace them within one year, or in thirty days if the systems were causing obvious and significant pollution. Owners were also ordered to correct certain less serious defects found during the inspections. The Nelson Lake and Barber Lake studies also looked at the age of septic systems (see table 4). On Nelson Lake 29 percent of systems and on Barber Lake 43 percent of systems were more than forty years old or of unknown age.

Another investigation conducted in Michigan's lake-rich Charlevoix County, using data from the U.S. Census and the Health Department of Northwest Michigan, found that up to one-third of septic systems installed in the county from 1959 to 1984 had not been replaced. This means about 2,040 homes, or 35 percent, may now have septic systems that are much older than their expected life spans.[9]

While many of the older systems in the Sawyer County study were found to be working properly, and presumably some of the older Charlevoix County

Table 4. Age of systems, Sawyer County lake septic surveys

| | Number of Systems | |
	Nelson Lake	Barber Lake
Unknown age	9 (2%)	1 (1%)
Before 1970	32 (7%)	10 (12%)
1970–1979	23 (5%)	7 (8%)
1980–1989	67 (15%)	19 (22%)
1990–1999	172 (38%)	25 (29%)
2000–2010	150 (33%)	23 (27%)
Total	453	85

Source: Sawyer County (Wisconsin) Zoning and Conservation Department.

systems were as well, age does to some extent correlate with failure. More to the point, the recipe for failure is age combined with neglect, observes the University of Minnesota's Anderson. A longtime educator of homeowners and a trainer of septic system professionals, Anderson draws an analogy between septic systems and household mechanical equipment. "For planning purposes, I'm going to have to redo my furnace every twenty to twenty-five years," he says. "If I do the maintenance and have technicians replace parts as I go along, the central unit is still there and functioning. But if I just let it go, I might have to replace that furnace after ten or fifteen years. The same applies to septic systems. It's not that you can just stick them in the ground, walk away, and let them be. After probably twenty to twenty-five years, you're going to have to do something with your system. But that is particularly true if you haven't done the maintenance and management of it over time. Owners also need to consider that the amount of water they use, the amount of solids they introduce to the system, and other things they put down the drain will have impacts."

Responsible Care

Maintenance and management is where the capacity of septic systems to protect groundwater and lake water quality often breaks down. While septic systems are much more closely regulated than they were fifty years ago, owners generally are left to their devices in the day-to-day use and care of their systems. The EPA's septic system program lists best maintenance practices that are widely promoted through county and local governments and by professional septic system service companies. They include:

Regular Pumping and Inspection

Solids slowly accumulate at the tank bottom and need to be pumped out periodically. At that time, the service provider should inspect the system for leaks or other defects and recommend repairs as needed.

Efficient Use of Water

Nearly all water used in a household ends up in the septic system. Too much water—from multiple laundry loads, long showers, hot tubs, water softeners, or leaking toilets or faucets—can overload the system, hinder treatment, and lead to failure. Water-efficient toilets and washing machines, faucet aerators, low-flow showerheads, and general water-conserving habits will help the septic system perform effectively and extend its life.

Proper Handling of Waste

The septic tank is not a catchall. Nothing should be flushed down the toilet except the three *p*'s: poo, pee, and (toilet) paper. A garbage disposal on a septic system is generally a poor idea—ground-up particles will fill the tank more quickly and escape the tank to clog the drainfield. Chemicals put down the toilet or sink, like drain openers, solvents, and harsh cleaners, can kill the helpful bacteria that break down waste in the septic system. Cooking oil and coffee grounds should be put in the trash.

Drainfield Maintenance

The drainfield does the real work of treatment and must be protected from soil compaction and flooding. Roof drains, sump pumps, and other rain-water drainage systems should be directed away from the drainfield. Vehicles should never be parked on the field; even foot traffic across it should be limited. Trees should not be planted near the drainfield, as root intrusion can disrupt treatment.

Safeguards through Regulation

Although this advice is widely available, compliance is uneven at best. The backstop against widespread mismanagement consists of various forms of state and local regulation. Four types are the most common.

System Inspection at the Time of Property Transfer

This approach aims to make sure that septic system defects are identified and corrected before the property passes from one owner to another. A local health department staff member or a private contractor performs the inspection according to a specified protocol. If the system is found to be deficient or failing, the property seller and buyer must agree on a plan to repair or replace the system. Where this kind of regulation does not exist, many systems are inspected anyway because lenders require it as a condition of granting a mortgage. Either way, a potential weakness is that properties transferred within families or through straight cash transactions can bypass the inspection.

Mandatory Pumping

These rules aim to make sure every system gets the most basic item of maintenance: the periodic removal of solids from the septic tank. Typically, the town or county government sends the homeowner a notice when pumping is due; the owner then must hire a licensed septic service contractor to pump the tank. These regulations can miss older systems that are not in the town or county records.

Mandatory Inspection

The inspections are typically required every five years and may also include pumping. Anderson notes that to be effective, the inspection needs to cover every component, including the septic tank, effluent screens, and pumps (if any). It also should include an inspection of the drainfield. The soil should be examined at the first inspection, and later on if there is reason to believe an important condition, such as the level of the water table, has changed. The owners must correct any defects found within a specified time.

Septic System Maintenance District

This is the most effective form of regulation but also the most complex and the least common. A special district is formed to actively manage all the septic systems around a lake or in a community. Property owners are assessed fees in return for receiving regular system inspection, maintenance, and repair. In essence, responsibility for system care shifts from the homeowners to the local authority.

Otter Tail Lake Project

Minnesota's Otter Tail Water Management District is an excellent example of the maintenance district approach. The district was formed in 1984 under state law as a way to ensure the proper, cost-effective treatment of wastewater in a fifty-five-square-mile area around six lakes. Initially the district encompassed about 1,200 homes, seasonal cabins, and businesses; today it includes roughly 1,850 dwellings connected to septic systems. Upgrades were needed because many of these systems were on small lakeshore lots and in sandy soils through which pollutants could easily migrate to the lakes. Many septic systems had been installed too close to the lakes and too deep relative to the lake's elevation, leaving inadequate separation from the water table.

Six lakes were included in the district because to varying degrees they interconnected. Besides Otter Tail, the district encompasses Blanche, Walker, Long, Round, and Deer lakes, and part of the Otter Tail River North. Septic systems were the primary but not the only source of pollution in the lakes, according to former district administrator Mann. A few farmers on creeks feeding into the lakes were encouraged to improve their practices—in one case Mann persuaded a farmer to erect fences to keep his cattle from entering a lake feeder stream. "And people who think they need to have a lawn that looks like a golfing green, they fertilize the heck out of it," Mann observes. "Over the years, we tried to educate people that with our permeable soil, you can't be dumping all that fertilizer on your yard. It's going to end up in the lake in a real quick time. The Otter Tail Lakeshore Property Owners Association was very good about reminding people that we all have to work to keep things nice here."

The main targets, though, were the inadequate and failed septic systems. "A few people had taken it upon themselves to fix their systems because they knew they were bad," Mann recalls. In addition, in the late 1970s, as discussions about district formation began, Otter Tail County was starting to inspect and condemn deficient systems on some of the lakes.

In its initial stages, the district inspected every system around the six lakes. Functioning systems were left in place. In the end, with help from an EPA grant, about 850 properties were upgraded. Of these, 590 new systems served individual homes and cabins, and sixteen cluster systems served a total of 260 properties. Cluster systems serve groups of homes; each home in the cluster discharges to septic tank, and the effluent from all those tanks is pumped through a sewer line to a single large drainfield at a distance from the lakeshore. "A lot of the cabins did not have room on the property for a drainfield," says Mann. "That was the reason for putting in the clusters. We pumped everything back off the lake, beyond the roads, to get the wastewater away from the lake completely."

The initial wave of replacements left some questionable systems still in place because under Minnesota Pollution Control Agency regulations, they did not qualify for the federal funding. But the district later started its own program to bring those systems up to code. "After four or five years, we had brought pretty much everybody into compliance that we knew of," Mann says. "Of course, as you go along there are always some that fall under the radar. They cropped up, and as we found them, we made sure we got them into compliance. Anything that was installed over the next thirty years obviously was built according to the code."

The system replacements were only the first step; the key to the district's long-term effectiveness is a two-tiered management program. Residents and businesses can choose an active or a lower-cost passive maintenance plan.

In the active plan, the district maintains the system; an annual fee covers regular pumping, inspection, and basic repairs. The owner pays for changes beyond the basics or for damage to the system caused by negligence, such as excessive use of water. All cluster systems are on the active plan.

Under the passive plan, the septic system is under the district's supervision, but the property owner is responsible for all maintenance and repairs. The district inspects the system (every two years for permanent residents

Phosphorus in Ottertail Lake

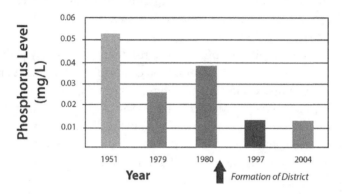

(Eric Roell; data from the University of Minnesota)

Water Clarity in Otter Tail Lake

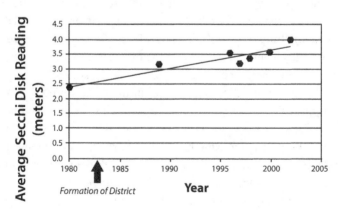

(Eric Roell; data from the University of Minnesota)

and every three years for seasonal properties) to assess septic tank, component, and drainfield conditions. Homeowners receive notices when system
pumping is due, and they send in a reply card when pumping is done. They
also receive information about best system management practices. Owners
have the option to switch to the active program, and since 1984 about thirty
have done so. Those on the active program may not switch to the passive.

By any measure, the program has been a success. Since 1986, only forty
septic systems in the district have been replaced, a failure rate of 2.4 percent. More telling is the effect on the lakes. Water clarity as measured by
Secchi disk readings has improved dramatically.[10] The preceding graphs
show the results for Otter Tail Lake, the largest in the district, which saw a
significant decrease in phosphorus levels. Similar data exist for the other
five lakes in the district.

The difference is readily apparent to lake residents. "Absolutely people
could see the improvement," Mann says. "You could go out in summer and
in ten feet of water you could see the bottom. Obviously, you're going to
have algae blooms—that's just Mother Nature at work—but by and large the
lakes definitely cleaned up. There's no doubt about that." The change is especially impressive in light of the intensive development on the shores: "There
is very high density on this lake. It's wall-to-wall cabins; every seventy-five
to a hundred feet there's a cabin."

Reflecting on the district's history, Mann notes that some $5.5 million in
federal funding under the Clean Water Act made the system replacements
and cluster system construction possible. "Otherwise it never would have
happened," he says. Cost would be a potential barrier to similar projects
today, he believes. Even with funding at hand, the idea of forming the district met with resistance. "We had numerous people around the lake who
didn't want anything to do with it," says Mann. "They were very adamant;
they were totally against it because of the extra layer of government. They
didn't think there should be any funding for private systems."

He strikes an optimistic note for the future: "The younger generation
coming up, they're much more prone to say, 'Hey, we've got to take care
of where we're at, or we're not going to have anything in twenty years.' The
old guard is kind of gone. Now people are so much more green-minded.
They're taking care of their systems better. There's better education by the
MPCA and the lakeshore association to really drive the point home that
you've got to take care of your septic system. Pump it religiously and you'll
negate a lot of problems."

Tempered Optimism

Beyond the Otter Tail district's example, there is room for a brighter outlook on septic systems. Significantly, the septic system maintenance industry has become more professional, and service contractors often are key sources of advice to homeowners on how to take care of their systems. Many offer regular maintenance plans and include education as an essential component of their service. Along with septic tank pumping, they inspect systems and alert property owners to issues that need attention, helping them head off failures and extend system life.

Meanwhile, slowly yet steadily, properties change hands and systems are inspected. As old, failing systems are discovered, they are replaced with systems built to current codes. In addition, septic system technology is improving. For example, there are pressure distribution systems that spread septic tank effluent more evenly across the drainfield trenches. Aerobic treatment units function like miniature wastewater treatment plants and produce much cleaner septic tank effluent than basic systems. These units typically are installed only in areas where regulators determine that a traditional system would not adequately protect the environment; they can also help rehabilitate drainfields that have failed.

In all types of septic systems, some states and localities require additional protective features such as:

- Effluent screens that remove larger particles from septic tank wastewater, helping to enhance drainfield performance and extend life.
- Inspection ports that make it easy to check drainfield trenches for ponding of water, a potential sign of impending failure.
- Ground-level lids that enable service contractors to access septic tanks for pumping and inspection without having to dig them up.
- High-water alarms that alert homeowners when the water level in the septic tank is rising, so that they can call a service professional for help before the system backs up.

Sawyer County's Eric Wellauer shares the optimistic outlook but tempers it with caution. Experience tells him that modern systems, cared for properly, have minimal impacts on groundwater and the lakes. "For the most part, I think they are doing what they are supposed to do," he says. "However, there probably are systems where a soil tester made the wrong determination, or the system was installed incorrectly."

Then there is the issue of older systems: "If they don't meet the current code, I think there are detrimental effects to the environment, whether it's drinking water from the wells, or surface water, or the organisms living in the lakes." He also sees homeowners' attitudes toward their systems as a concern in that people in his county seem less receptive than in the past to septic system education and to fixing system problems. "People spend half a million dollars or more on lake homes—they want the view and they like the quality and clarity of the lakes." At the same time, he observes, many balk at investing several thousand dollars for septic system replacement.

Anderson adds, "Around my home I see resistance to the county program of septic system inspection and pumping. Yet people have expectations that the lake quality needs to be the same or better. It is a continuing education proposition. I see a lot of big-city people moving or buying property around me, and they have no concept about why some of these things are important."

Taking Responsibility

Sara Heger, at the University of Minnesota Water Resources Center, says some property owners who resist inspections take refuge behind the concept of private property rights, a position she rejects.

Our water is all connected, so at the end of the day, what we're doing on our property is affecting the water in the lakes or the water in the ground. The thing to keep in mind is that when we use the land, however we use it, we can negatively affect water quality. Everyone wants to blame someone else for why the water isn't the way we want it to be. I believe we all need to do better to protect the lakes.

To me, the whole perspective is that if I am a rural landowner, my septic system is my responsibility. To say "I don't want to have to pay for a new one," that is not right. I have lived in Minneapolis for twenty years. I pay a hundred dollars a month for sewer and water, so I've paid as much as a well and septic would cost. Whether you live in an urban or rural area, part of the cost of home ownership is water and wastewater. If your septic system isn't properly treating, it's your responsibility. If we want septic systems to be viewed for what they are, which is a part of our permanent wastewater infrastructure, we need to hold them accountable on some level. So having management programs where every septic system is looked at every three years at least, that's where I believe we need to get to in order to protect our water.

~~~~~~

CHAPTER 9

In the Wake

All of a sudden on this beautiful, peaceful lake, a wake boat starts up.
I didn't even know what wake boats were. The boat's stereo was just
booming; you could feel the vibration. They stirred up the whole lake.
They didn't care where they went, and they created these huge waves.
I almost thought I was in a time warp. It was a beautiful day, and all of
a sudden this thing just blasted in. It was a complete bully on the lake.
I said, 'Wow, why are they allowing this boat on the lake?' It consumed
everything. It seemed so unfair.

— Interview with lake visitor, June 2020

Attend almost any lake association meeting or gathering of lake advocates
and you are likely to hear about wake boats. These large boats, typically
twenty to twenty-six feet long, with seating for multiple passengers and
inboard engines from about 250 to more than 600 horsepower, generate
large waves on which wakeboarders and wakesurfers can perform tricks.
To say these boats cause consternation among lake residents is to under-
state the case severely.

Conflicts between motorized and quiet water sports have existed for
decades. Waterjet-propelled personal watercraft hit the market in the 1970s;
many people still consider them annoying. Many are also irritated by speedboats and by waterskiing or tubing. Some lake groups make peace by institut-
ing courtesy codes and designating quiet hours in the morning and evening.
Wake boats appear to be another matter: critics say that when used irre-
sponsibly and even as intended, they have potential to cause harm to property, to other lake users, and to lake ecosystems. The waves' side effects, say

critics, include erosion of shorelines, damage to piers and moored boats, disruption of wildlife habitat, and hazards to children on beaches and to people fishing, canoeing, kayaking, and paddleboarding. Many wake boats have high-wattage music systems that users sometimes play at volumes others consider obnoxious. Initiatives are underway in various states and localities to regulate these boats. Many people would prefer to see them simply disappear.

Lake advocacy organizations are taking notice. Michigan Waterfront Alliance board member Scott Brown observes, "A dramatic upsurge in the number of wake boats operating on Michigan's inland waters has resulted in a significant escalation in reports of overturned kayaks, swamped fishing boats, and damage to both developed and natural shorelines. The increasingly alarming situation has not only elicited the intense anger of affected waterfront property owners, paddle boat enthusiasts, and fishermen but it has also helped attract the attention of local, county, and state government officials."[1] In the opinion of Clifford H. Bloom, attorney for the Michigan Lakes and Streams Association, "Wake boats are not simply a different type of boat, and the problems they create are not just a matter of degree. The problems caused by wake boats are geometrically worse than conventional speed boats."[2]

Michael Engleson, executive director of Wisconsin Lakes, the statewide lake association, notes that wake boats are proliferating at a time when a series of wet years have raised lake water levels, and when boat traffic increased as people sought safe recreation during the COVID-19 pandemic. "It is becoming the major issue on our lakes right now, because it reaches so many different pieces of the problem," Engleson says. "The wake issue cuts across water quality; it cuts across habitat, shoreland preservation, safety issues, and aquatic invasive species. It's a level of impact on the waters that we have not seen before."

Gaining Popularity

Sales of wake boats are growing. An executive for boat manufacturer Master-Craft credited wakesurfing for helping revive the ski-boat industry after the Great Recession of 2008–9.[3] The National Marine Manufacturers Association (NMMA) estimated that thirteen thousand wake boats were sold in 2020, up 20 percent over the previous year.[4] Wakeboarding and its cousin wakesurfing are arguably the fastest-growing of motorized water sports. Both are undeniably fun; wakeboarders especially can make spectacular soaring leaps, spins, flips, and twists as they cross back and forth over the boat wake. Riders' feet are strapped into boots attached to the board; as in waterskiing the rider holds on to a rope, typically sixty-five to eighty-five feet long. Wakesurfers' feet remain free; they can move about on the board in the manner of ocean surfers, although the board is much smaller. They start by holding onto a rope just ten to twenty feet long; when up to speed, they throw the rope forward into the boat and cruise along hands-free on the wake on one side of the boat. The tricks are less dazzling than in wakeboarding, although expert surfers can execute a variety of spins and leaps. For wakeboarding, the boat typically travels at fifteen to twenty-five miles per hour; for wakesurfing, eight to eleven miles per hour. Wakesurfing produces the largest wakes; in on-water boat testing, waves have been measured at 4.2 to 4.6 feet from trough to crest at surfing distance behind the boat (the amplitude quickly decreases farther back).[5]

Jeff Forester, executive director of Minnesota Lakes and Rivers Advocates, readily understands why wake sports are popular.

People who are older can learn to wakesurf. It's not very hard. You're not going really fast, so if you fall you don't pay the price as you would if

Wakesurfers ride on enhanced wakes created by high-powered boats. (Adobe Stock Images)

waterskiing. You can have the whole family in the boat. They've got refrig-
erators and sound systems. They have all the bells and whistles. On the
other hand, people who live on the lake see the impact of the wakes on the
infrastructure, the aquatic plant beds, water clarity, shore nesting birds
like loons, and shore erosion, and they are also really passionate. These
are really powerful watercraft, and they can change a lot in a very short
period of time.

Making the Wake

The sport itself is evolving. As with any new activity, wakeboarding and
wakesurfing come with the drive to innovate, and to a large extent that
involves improvements in the boats. Engineers are ever at work on designs
that create bigger and better wakes. Four factors work together to create
wakes suitable for boarding and surfing.[6]

Ballast

The wake springs mostly from water the boat displaces: more weight means
bigger wakes. Wake boats have ballast tanks that can be filled with water
weighing one to three thousand pounds; some up to six thousand pounds.
Pumps can fill or empty the tanks in a few minutes.

Water Flow

Each boat maker has its own wake-shaping system. Some use devices like
large trim tabs to deflect water down at a severe angle. Others use a blade,
wedge, or contoured plate to channel water from one side of the boat to
create a perfectly shaped wave on the other side. Some have remote devices
that let surfers adjust the shape and size of the wave while in action.

Hulls

Hull designers aim to produce waves as large and crisp as possible. Models
vary, but most are deep-V-shaped hulls with a hard keel at the bow that narrows at the stern corners. This angled hull surface helps shape long and
powerful waves.

Propulsion

It takes a great deal of power to carry thousands of pounds of ballast and
move the displaced water to make waves. Types of engines vary, but boats
for wakesurfing, for example, generally require at least 250 horsepower.

The National Marine Manufacturers Association gave its 2020 Innovation Award to the Waketoon boat from Avalon, a luxury pontoon boat (priced over $150,000) that can also be used for wakesurfing. The boat, driven by a 380-horsepower V8 engine, carries no ballast tanks; it creates large wakes by way of its pontoon design. Meanwhile, aftermarket accessories make it possible to convert a traditional runabout boat with an inboard engine into a wake boat, or to upgrade a wakeboarding boat for wakesurfing. For example, as of fall 2020, a Go Surf Assist conversion system from Wakemakers cost less than $5,000; other wake-enhancing accessories were available at prices as low as about $1,100.

Wake boat manufacturers proudly tout the size and quality of the waves their boats produce; their advertising reveals a stiff competition for the biggest and best-configured wake:

- "The Pavati AL24 and AL26 wakesurfing boats throw the best wakesurf waves on earth, courtesy of . . . nearly 6,000 pounds of hard-tank ballast and our revolutionary Rip Tide Surf System. The AL Series wakeboats . . . engineered the ultimate wakesurf wave with more height, length and push than any other wave ever produced. The only real competitor to a Pavati wave is an ocean surf wave."[7]
- "The revolutionary [Gigawave] is an electric powered watercraft boasting the largest wave ever created. . . . Sculpted by the wave manipulation

Wake boats use ballast, special hull designs, and other features to create large waves. (Adobe Stock Images)

system and given its size from the largest displacement wake surfing hull ever built, the continuous, head-height wave matches the size and feel of ocean waves. It's big, powerful, and clean; delivering a massive barrel that will take the sport to new levels."[8]

- "[With the ZS232] you will feel like Zeus unleashing a lightning bolt as you dial in the perfect wave to suit your style. The Tapered V-Hull helps you soar to new heights when you launch from the powerful surf wave you create."[9]

Boat Industry Perspectives

For all the sport's pleasures, a key question is whether wake riders and their boat drivers appreciate what the wakes have wrought on the lakes and their shorelines. Boat manufacturers, other equipment suppliers, and many wake sport enthusiasts are well aware of concerns about their impacts. To help fight off multiple attempts at local and statewide regulation, and to address property owners' and lake users' concerns, the Water Sports Industry Association (WSIA) commissioned a Towed Water Sports Wave Energy Study, conducted in spring 2015 on a chain of lakes in Orlando, Florida.[10] The study measured shallow- and deep-water wave energy from a wake boat carrying 4,250 pounds of ballast and with a total weight of five tons; it also looked at the effects of wind-driven waves. The study concluded that "in most conditions, wakesurfing and wakeboarding are far less destructive than naturally occurring waves." It also reported that waves from wake boats operated at least two hundred feet from shore "do not carry enough energy to have a significant impact on most shorelines or on properly maintained docks and other man-made structures," and that the wave heights from wake boats drop in size faster than those from other recreational boats. With this study in hand, the WSIA launched a "Wake Responsibly" campaign that encourages wake-boaters to be considerate and to operate in ways that minimize damage. It asked sport participants to make a pledge to limit repetitive passes over the same area, keep music at a reasonable level, and stay at least two hundred feet away from docks and beaches.

Larry Meddock, recently retired as WSIA chairman, blames much of the backlash against wake boats on irresponsible operators. He recalls years ago when stores sold a plaque that boat owners could place near the helm, stating: *I am the captain of this ship and I will do as I damn well please.*

You can't buy that plaque anymore, but unfortunately that attitude prevails today as much as it did thirty or forty years ago. That is troublesome, and unfortunately, we as an industry absolutely accept the fact that we've got some knuckleheads, and that we have an obligation to educate captains in the responsibility that they take on when they get behind the helm of a boat. When you're wakesurfing, and you're going somewhere between eight and eleven miles per hour, the captain thinks, "I'm going so slow, I'm surely being responsible. I can't be bothering anybody. I'm just chugging along out here not bothering a soul." And he's clueless of what his wake is doing behind him. And so we've got to bring that to the forefront. . . . We've got work to do.

On the government side, the Wisconsin and Minnesota DNRs have partnered on an education campaign called "Own Your Wake"; it encourages users of all motorized watercraft to be good neighbors and reduce big wakes that could endanger other boaters, paddlers, anglers, swimmers, and shoreline property owners. Launched in July 2019, the campaign was driven by a growing number of calls and complaints in both states about excessive wakes. Minnesota's "Own Your Wake" webpage advises boaters to stay at least two hundred feet from shore and suggests that boats creating

The emblem of the Minnesota DNR "Own Your Wake" initiative. (Courtesy of Minnesota Department of Natural Resources)

"an artificial wake" may need to stay even farther offshore to limit the impact.[11]

Meanwhile, the Midwest Wakesurf Association, also based in Minnesota, advocates for keeping lakes and rivers open to the sport while helping educate wakesurfers on safe and responsible boating practices.[12]

Conflicts Growing

Many lake residents, lake association leaders, and quiet sports enthusiasts believe those educational measures are not nearly enough and strongly take issue with the WSIA study findings. Jeff Forester says he "would be surprised if the data supports" the industry's guidance of operating as near as two hundred feet from shore. He also believes the WSIA study inherently lacks credibility: "Big tobacco did all the research into lung cancer, and they cooked the books. So now nobody believes industry reports. The research has to be independent or it just doesn't have any meaning."

Chuck Becker, a resident of five-hundred-acre Big Sugar Bush Lake near Detroit Lakes in west central Minnesota, asserts that wake boats can significantly change a lake's character. As a founder of the grassroots group SafeWakes for Minnesota Lakes, Becker has spoken to numerous lake residents about wake boats and their impacts.[13] "I talk to a lot of elderly people who have had their lake places for decades," he says. "These boats have totally changed their quality of life. They can no longer enjoy their lakeshores. They can't let their grandchildren play on the beach because they get knocked over. One gentleman said he took his pontoon boat off the lake because he thought he could no longer safely take his grandchildren out. A woman on Lake Minnetonka has been there since 1953; she grew up on the lake. She was virtually in tears."

Dan Butkus, whose family owns property on 398-acre Squash Lake in the heart of northern Wisconsin's lake country, notes that with lake water levels near historic highs, large boat wakes threaten severe shoreline erosion. "We started the year [2020] with about eight wake boats; now we have thirteen," said Butkus, treasurer of the Squash Lake Protection and Rehabilitation District and a board member of Wisconsin Lakes. "Even though there is a code of conduct, or whatever the manufacturers call it, people will zoom by with their full wakes. I saw one wave travel a good thousand feet; it crashed into the shore across the bay. It was just this one giant rolling wave. They come in, and they're usually fairly broad as well as high, so a lot of volume of water and power comes with them."

He recalls another incident in which he and others on a pontoon boat watched a wave from a wake boat roll up onto an area of relatively flat shoreline. "There were kids on the beach, and there were families," he says. "We saw this wave roll toward them. They all went running, and the wave went inland like a tsunami. It was almost like slow motion, and it traveled at least fifteen feet up onto the shore. People were scattered, and there were inflated toys all over the place, stuff just tossed around. Nobody was hurt, thankfully." He estimates the boat that made the wake passed about two hundred feet from the shore.

On some lakes, a secondary effect of wake boat waves is property own-ers installing rock riprap or timbers to prevent shoreline erosion. This conflicts with initiatives to encourage natural shorelines and the ecologi-cal benefits they bring. Wake boats moved from lake to lake also threaten to spread invasive species contained in ballast water that remains after the tanks have been pumped out. Safety is another issue. Butkus recalls a near tragedy on Squash Lake: "Because wake boats ride bow up, with ballast pushing the back end down, it's hard for the driver to see over the bow. The chairman of our lake district was out with his grandson last year in a canoe, and along came a wake boat, heading right toward them." Fortunately, the wakeboarder saw what was about to happen and dropped into the water, so that the boat driver would stop and turn around. A more serious incident took place on July 4, 2020, on Jewett Lake near Fergus Falls in Minnesota's Otter Tail County. Three fishermen were forced to leap into the water for safety before their boat was struck on the side by a wake boat carrying three passengers and with a wakesurfer behind. No one was injured.[14]

Relying on Science

Proponents of wake boat restrictions are seeking data on the impact of large wakes to support the creation of a sound regulatory scheme. They also hold out hope that scientific facts about the boats' effects could help persuade wake boat enthusiasts to operate more responsibly.

WSIA's Meddock is skeptical about regulation of wake boats but does not reject it out of hand. "We believe that our society has more regulations already than it knows how to process," he says. "Every time we have ever been in front of a legislative body, whether it be in Wisconsin, Minnesota, or Michigan or some other state . . . whenever we ask if we can give educa-tion a try before we regulate, without question they will agree to that. Which to me underscores that our society today is overregulated, and most people we talk to don't want to see more regulations."

At the same time, he says the industry could support responsible regulation based on science. Meddock strongly defends WSIA's 2015 study.

The bottom line is, the guys we had do this were scientists and engineers out of MIT, and they will stand behind their data to this day. They will take on any peer review you want to throw at them, and they are prepared to defend the science—because the science is the science and the data is real. The problem we have is these issues from property owners and others who claim we are destroying their lakes but have no science behind their accusations. They simply throw anecdotal comments out and demand that there be change. You show us the science behind your request and the WSIA and the NMMA will take a really hard look at it and adjust as appropriate. We're all about compromise. We think a collaborative effort with industry and environmentalists will provide a much stronger, stable platform to work from. When the science comes out, that is what we should all appropriately respond to.

He states, for example, that the industry could support limits on wake-boating based on lake size, depth, and configuration where justified by data.

A substantial amount of data beyond the WSIA study has been and is being generated. Two university studies on wake boats have been completed on lakes in the Canadian province of Quebec. One study, led by the Université du Québec à Montréal, looked at the impacts of oversize wakes on two important recreational lakes. The study found that a wake boat causes "considerable impact on the shore when it passes 100 m [330 feet] to the shore, and all passages within 300 m [990 feet] significantly add energy to the waves naturally present."[15] The researchers observed that waves created for wakesurfing have the greatest impact, 1.7 times greater than from a boat of normal displacement. They concluded that to eliminate impacts beyond natural wave action and to avoid shoreline erosion, regulations should require wake boats to operate at least three hundred meters from shore.

The other Quebec study, conducted by Université Laval on two lakes, looked at the impact on the water column and bottom sediment from the propulsion systems of wake boats with more than 350 horsepower.[16] The researchers concluded that under the conditions studied, wakesurfing and wakeboard boats had potential to remobilize bottom sediments in depths up to five meters (sixteen feet) for more than one minute. Referring to the above-mentioned study, they stated that to prevent shoreline erosion, and to avoid the suspension of sediments and thus the availability of phosphorus in the water column, wake boats should be limited to lake areas at least six hundred meters (1,980 feet) wide and five meters deep. The study did not account for the possible cumulative effects of multiple boats passing over the same places.

One Lake's Experience

Closer to home, a three-year scientific wake boat impact study is in progress on North Lake in the glacial lake country of southeastern Wisconsin. Its results could reverberate across that state and far beyond, since the wake boat challenges there are shared by numerous lakes in the Upper Midwest and elsewhere. The study, under contract with the North Lake Management District, is being conducted by Carroll University, the Southeastern Wisconsin Regional Planning Commission, and Terra Vigilis, a critical infrastructure and security company equipped to perform environmental studies using drones. The study is examining wave dynamics and the effects of boat wakes and propeller downwash on water clarity and chemistry, aquatic plant life, and bottom sediment. The study grew out of complaints the district received about wake boats starting about six years ago, according to Timothy Tyre, cochair of the lake district's stewardship committee and an aviation science faculty member at Carroll University.

North Lake actually consists of two lakes—a 340-acre main lake and a 108-acre lake connected to it by a narrows. Controversy began when the first wake boats appeared on the lake in 2015. People complained to lake district leaders about large waves damaging the shoreline, knocking boats off lifts, and causing various other problems. In response, the district sent a survey to all lakefront property owners to gather their opinions. The replies came back expressing strong opposition to wake boats, but wake boat owners criticized the survey as poorly designed and thus invalid, Tyre recalls.

There were twenty-three wake boats on North Lake in summer 2020. Tyre says their effects on the shorelines are substantial. The lakes have a maximum depth of seventy-eight feet, but around the perimeter of each is a shelf with water three to four feet deep that extends out a hundred feet or more, at which point the bottom drops off at about a 45-degree angle. Tyre observes, "Propagating waves, particularly on the small lake, is a real issue, because when the energy finally gets to that shelf, it touches the bottom, and a wave that seems to have disappeared suddenly reappears. It's like a tsunami. All that subsurface energy is pushing the water back up, and pretty soon we've got whitecaps hitting the shore."

The lake is patrolled by two municipal law enforcement agencies and by DNR wardens, but they have no authority over wake boats beyond enforcing local and statewide regulations against such activities as excessive speed, reckless behavior, and operating less than one hundred feet from shorelines, docks, and piers at faster than no-wake speed. In 2017, the district established a lake usage committee with members representing sailors, water ski boaters, personal watercraft owners, wake boat enthusiasts, and others. The group met monthly for a year. During that time, Tyre, a group member, developed a science-based report describing the lake, its features, boating activities on the lake, boat wave physics, and boating regulations and their enforcement. Meanwhile, the stewardship committee developed a set of voluntary boating courtesy and safety guidelines. In part it encouraged wake-boaters to stay toward the middle of the lake, and it established slow no-wake hours on Saturdays and Sundays after 6 p.m. to allow the lake to rest. "What we're really trying to do," says Tyre, "is change the culture of the lake voluntarily. The compelling variable in changing the culture is hard data."

The first phase of the North Lake study was completed in summer 2020. A key component was the use of drones to image the lake and the sediment plumes generated by wake boats. According to Tyre,

> We can aerially image the differences in wave propagation by boat type. These views are very, very compelling. We've also put underwater drones to work so we can image the subsurface impact. The wave energy goes up [in the water column], and of course it also goes down. When it moves out from the boat and finally touches the bottom, it stirs up the bottom, and it's creating big plumes of sediment that redeposit in various places. We wanted to figure out what's in those deposits, and we wanted to look at what is stirring up from the bottom. So we took core samples, and we used the university lab's ion chromatography technology to nail down what is in the dissolved sediment in the lake.

The study also looked at sediment deposition: "At the beginning of summer, we did subsurface drone photography of all the weeds that were being stimulated by the plumes, because they carry a lot of nitrogen and phosphorus."

After the study's first phase, "What we can say so far with a high degree of confidence is that there is a fundamental difference between the waves that are propagated by ski boats, personal watercraft, pontoon boats, and wakeboard boats," Tyre says. In particular, the data demonstrated that wake boat waves have greater amplitude and carry more subsurface energy than those from the other watercraft. The preliminary data also showed differences in sedimentation near shorelines. "With pontoons and personal watercraft, you don't see much scrubbing on the bottom," Tyre says. "You don't see as much turbulence and sediment plume activity as you do from wakeboard boats. The wakeboard boats, because of the depth of the plume and the energy that's scratching the bottom, so to speak, they create big plumes that run along the shoreline and then jump over shallow-water reefs."

One such reef, under six to eight inches of water, lies at the narrows that separates the two lakes. Drone photography showed a sand plume from a wake boat wave following the shoreline, meeting and jumping over the reef, and depositing sediment on the other side. "That is a wakeboard-boat-specific phenomenon," says Tyre. "At the end of summer, we put our sub-surface drone down, and over an area probably two hundred feet long and sixty to eighty feet wide, in about fifteen feet of water, all the plants were covered with an inch or more of what looked like droppings from volcanic ash. It was very dramatic."

In February 2020, the lake district received three grants from the Wisconsin DNR totaling $30,000 to continue and expand the wake boat study. The 2021 phase of the investigation looked at the depth of downwash from the wake boats' propulsion systems, its related impacts on shorelines and sediment redeposition, and the water quality effects of suspended solids and phosphorus stirred up from the bottom. To aid in assessing propulsion system impacts, Terra Vigilis designed a device able to probe as deep as twenty-five feet, with cameras spaced every five feet to observe the behavior of streamers tethered to a line. The study found that propeller downwash from a high-powered, latest-technology wake boat in wakesurf mode disturbed the lake bottom to depths up to twenty feet, both at startups and when running past markers, Tyre reports. The researchers also acquired up-to-date bathymetric data on the lake and then used a computer model to simulate a wake boat moving across the surface and producing a fifteen- to twenty-foot downwash plume. The aim was to assess the impacts on the bottom topography anywhere in the lake. "It's very compelling when you look at it," says Tyre.

This phase also measured the shoreline impact of waves from the four types of craft most often used on the lake at two hundred, three hundred, and four hundred feet from shorelines. Pontoons, personal watercraft, and fishing boats showed minimal impact. For wake boats in the wakesurf mode, says Tyre, a two-hundred-foot distance did not appear to be adequate to protect shorelines. Impacts even at three hundred feet appeared "fairly significant"; impacts did begin to mitigate at four hundred feet. As of September 2021, the study data had been turned over to hydrologists for analysis. In addition, the district intends to examine the impact of wakes on phosphorus resuspension, and on dissolved oxygen in deeper areas of the lake.

Crowdfunded Research

Meanwhile, in autumn 2020, the University of Minnesota's St. Anthony Falls Laboratory conducted a study on Lake Independence, west of the Twin Cities, to measure how boat-generated waves and propulsion jets affect lake shorelines and bottoms, and how extreme waves change those impacts. The study included wake boats and direct-drive and stern-drive ski boats. The researchers used sensors and cameras above and below the surface to gather data from boats operated under different sets of conditions, including various water depths and distances from shore.

The study's proponents included the SafeWakes group, Minnesota Lakes and Rivers Advocates, the Minnesota Coalition of Lake Associations, and the Gull Chain of Lakes Association. Unable to secure funding through state government sources, the group turned to crowdfunding through the University of Minnesota Foundation, seeking to raise $94,000 to cover the first year of research. Likely reflecting the depth of concern about wake-boating impacts, the appeal quickly surpassed its goal, raising $143,000 by mid-December 2020 from numerous lake associations, individual donors, one local marina, and even statewide lake advocacy associations from Michigan and New Hampshire.

The study results were released in late January 2022 after peer review.[17] The wakesurf boats tested were a twenty-one-foot Malibu VLX Wakesetter and a twenty-five-foot MalibuMXZ Wakesetter. The other boats were a twenty-one-foot Larson LXI 210 and a twenty-foot Malibu Response X commonly used for cruising, tubing, waterskiing, and wakeboarding.

The boats were driven at distances of 225 feet, 325 feet, 425 feet, and 625 feet from shore. All boats were operated to plane on the water's surface (about 20 mph) and to run in typical wakesurf mode (about 10 mph and plowing, with the bow high and the hull displacing a large amount of water to produce the largest waves). Wave height (trough to crest) was measured for each boat and later analyzed to calculate total wave energy (the ability of the wave to do work) and maximum wave power (the rate of energy transfer). The study found that waves produced by wakesurf boats in the wakesurf mode were measurably greater than those of the other boats under all three criteria.

Specifically, when comparing the boats under typical operating conditions, at a distance of one hundred feet from the boat, the wakesurf boats produced maximum wave heights that were about two to three times larger, total wave energies that were six to nine times larger, and maximum wave powers that were six to twelve times larger than the non-wakesurf boats. The study also found that a wakesurf boat in the plowing mode would have to operate more than five hundred feet from shore for the waves to subside to the levels of a non-wakesuft boat two hundred feet from shore.

The study did not project how the larger and more powerful wakesurfing waves would affect shoreline erosion or shoreline structures, nor did it assess the impact of propeller wash on bottom sediments. Instead, it stated that the results would support further research on those impacts and help inform recommendations related to boating policies or legislation.

There was urgency behind the study, according to Chuck Becker of SafeWakes, because in early 2020 the Minnesota legislature debated a bill, supported by the boating industry, that would have required wake boats to operate at least two hundred feet from shore—in line with the industry's Wake Responsibly guidelines. Lake advocates pushed back hard. As SafeWakes cofounder JoAnn Syverson stated in a newspaper guest editorial in March 2020, "The one-size-fits-all 200-foot setback in the industry-sponsored legislation will not protect our lakes from wake boat damage."[18]

The legislation received a hearing before a committee in the Republican-majority state senate and was approved on a 7–5 vote, but it did not reach the floor. Still, Becker and other lake advocates say the battle is not over. "We firmly believe that the boating industry will be back . . . trying to get this through," says Becker, who testified against the legislation at the hearing. "The industry would like nothing better than for Minnesota, the Land of Ten Thousand Lakes, to pass this bill, because it would then become the standard bearer. Anytime some other jurisdiction would bring up wake boat regulations, they would say, 'Wait a second. The state of Minnesota—and who should know better than they—passed a bill at two hundred feet.'" In fact, a similar bill was introduced in the Minnesota legislature in early 2021; it did not advance.

Approaches to Regulation

Even with scientific data behind it, regulation is bound to be tricky, not to mention contentious. Attempts at regulation at the state and local levels have met with mixed success and have been vehemently opposed by the national boating industry, boat dealers, and wakeboarding and wakesurfing enthusiasts.

Michael Engleson, of Wisconsin Lakes, enumerates some of the obstacles to regulation. He notes first of all that in his and other states, the Public Trust Doctrine holds that waters should be freely accessible. He says, "People should be able to use the waters the way they desire, and the presumption is that we should allow that access and that use. So where we restrict that, we have to show that some sort of problem exists that infringes on the public trust rights of other users or is somehow damaging the resource."

That leaves the question of how and what to limit, and how any regulation would be enforced. Limiting wake boats to lakes of certain sizes would be difficult: a lake with seemingly suitable acreage might still be ill-suited because it is shallow or irregular in shape so that few large open spaces for wake-boating exist. Meanwhile, "If you create an ordinance that says no enhanced wakes closer than five hundred or six hundred feet from shore, that gets really hard to determine on a lake-by-lake basis," Engleson says. "Enforcement officers become uncomfortable making the claim because it's a question of whether they can prove that a boat was at the distance specified." As for regulating the size of a wake, he says, "Especially when it's well out onto the lake, if a citation is challenged, does it stand up in front of a judge? And how does the enforcement officer prove that the wake was too high? Even just a definition of what is an enhanced wake can be questionable. And then we have seen in other states that there is a vested business interest in keeping these boat sales going. So the preference would be to not have regulation."

Nevertheless, some local jurisdictions have put ordinances on the books. The Wisconsin city of Mequon, a Milwaukee suburb, regulates wake-boating on the Milwaukee River not by banning certain types of boats or limiting wake heights but by restricting certain uses of boats. The ordinance states:

No person may operate a boat in an artificially bow-high manner, in order to increase or enhance the boat's wake. Such prohibited operation shall include wake enhancement by use of ballast, mechanical hydrofoils, uneven loading or operation at transition speed. Transition speed means the speed at which the boat is operating at greater than slow-no-wake speed, but not fast enough so that the boat is on plane.[19]

Engleson observes, "What I like about the Mequon ordinance is that it not only bans the use of enhanced wakes, it also bans the use of most of the things that create them. So if your ballast tanks are full and the bow is way up in the air, then the violation is just doing that. It's not whether the wake is of a certain height. The presumption is that if you are using something that is engineered to create a higher wake, then you are in violation of the ordinance."

The Mequon ordinance does not address distances from shore because it applies only to a relatively narrow stretch of the river. An ordinance in the Town of Bass Lake in northwest Wisconsin's Sawyer County includes provisions similar to Mequon's but applies to lakes and therefore does specify distances: it prohibits the creation of an elevated wake (in excess of twenty-four inches) more than fifty feet long within seven hundred feet of the shore or shoreline structures.[20]

"As more and more lake communities begin to see the impact of these boats, I think we're going to continue to see attempts to pass ordinances like this," Engleson says. "In Wisconsin, those ordinances have to go through the scrutiny of the DNR, but there is the precedent of a couple that have already been accepted." He sees a potential role for lake districts in helping to manage wake-boating and recreational boating in general: "With the increase of enhanced-wake-creating watercraft, the current high water levels, and the amount of traffic on our waters, ordinances and local control are going to become very important. There are roles that lake districts can play, such as in getting a township to pass an ordinance on their behalf, and then managing the enforcement of it."

State-Level Initiatives

Regulations have been more challenging to enact at the state level, partly because it is difficult to write a single law to cover an enormous variety of lakes, and partly because the boating industry contests such laws vigorously. For example, in Vermont, a bill introduced in early 2019 would have limited wake boats to a speed of less than five miles per hour within two hundred feet of a shoreline, and would have prohibited wake boats from "plowing," defined as "intentionally operating a vessel with the bow elevated" on inland waters.[21] In response, an NMMA government relations bulletin declared that the proposed legislation amounted to a statewide ban of ballasted boats and stated, "Blocking passage of this bill is a top priority for NMMA and the Watersports Industry Association."[22] To date, the bill has not been enacted.

In New Hampshire, a Commission to Study Wake Boats met during 2019–20 to gather information on wake boat effects related to the spread of aquatic invasive species, shoreline erosion, impacts to private property, the safety of swimmers and other boaters, and the economic impact of recreational boating and the popularity of water sports. It included representatives from the state legislature, boating industry groups, wake boat enthusiasts, lake associations, shoreland property owners, and others. In the end, the group agreed on the merits of various wakeboarding safety regulations but not on any restrictions on wake-boat-related activities. The NMMA "applauded the report's conclusions and commended the commission for its comprehensive and fair and balanced analysis of the issue."[23]

In Oregon, in late 2019, the seven-thousand-member Willamette Riverkeeper organization petitioned the Department of State Lands to outlaw medium and large wakesurfing boats on a crowded stretch of the Lower Willamette River in Portland. The petition stated, "DSL action limiting wake surfing boats is necessary due to the large, powerful boats repeatedly and unacceptably posing risks to human life, health and safety; loss and damage to real property; irreparable loss and damage to natural resources and the environment."[24] The NMMA opposed this regulation, as well. However, in early 2022, both houses of the state legislature passed a bill banning wake-surfing on a twenty-nine-mile stretch of the river known as Newburg Pool. As of mid-March 2022, the bill was awaiting the governor's signature for final adoption.

Even local wake boat regulations often face boating industry opposition. In April 2019, "NMMA joined more than 250 stakeholders to help defeat a measure in Valley County, Idaho that would have placed punitive restrictions on wake boats," according to another NMMA bulletin. "The proposal would have disallowed waves greater than 24 inches within 1,000 feet from shore on seven popular lakes in southern Idaho. ... NMMA, the Personal Watercraft Industry Association (PWIA), and the Water Sports

Industry Association (WSIA) testified in opposition to the measure at two public hearings. . . . More than 3,100 advocates signed an online petition opposing the move, partly out of concern that the restrictions would have a negative impact on the county's economy."[25]

Role for Education

Until such time as workable regulations are adopted, harmony between wake-boaters and their detractors depends on boat operators understanding their potential impacts and observing proper courtesies and practices. To advance that cause, the WSIA promotes its Wake Responsibly campaign nationally and locally, through state departments of natural resources, municipal parks and recreation departments, boat manufacturers and dealers, lake and homeowner associations, and directly to boaters. It spreads the messages by way of social media, websites, email campaigns, promotional flyers, counter cards for dealers, boat ramp signs, word of mouth, and other channels. In addition, WSIA and boat dealers created a Wake Responsibly Compliance Exam with questions designed to make boaters aware of common boating and wakesurfing etiquette. The association says many boat manufacturers plan to integrate the exam with their new-customer education programs and boat delivery paperwork.

Forester, of Minnesota Lakes and Rivers Advocates, recommends going a step further by requiring operators of powered boats—not only wake boats—to be licensed. He observes that motorized boats are the only motor vehicles in his state that do not require licensing. He also notes that in a great many cases, the boats are not driven by the adults who buy them but by their sons or daughters. He believes that if families were to learn best boating practices, basic courtesy, ethics, and safety—perhaps with a unit on lake ecology in the mix—the behavior of wake-boaters would improve: "If the wakeboard boats were not causing problems, if they were being used out in the middle of large lakes, if people were being respectful and weren't going back and forth in front of the same shore, pounding it for hours, I think people would see that wake boats are a lot of fun, and they might not develop bad attitudes about them."

A fair question is whether the Wake Responsibly campaign's advisory of operating two hundred feet from shore might add a suggestion to stay as far from shore and as near to mid-lake as possible. Chris Bischoff, until 2021 the WSIA waterways access and government relations specialist, did not take issue with that concept. He noted that the recommended two-hundred-foot minimum is supported by the association's study, but added, "I'm not say-ing we couldn't adjust our language in the future. A person who's doing wakeboarding or wakesurfing obviously wants a bigger wake. A bigger wake is generated when the boat is in deeper water." And by extension, he said, educating boat captains to stay out deeper will enhance enjoyment of the sport while also protecting shorelines and minimizing conflicts with lake residents and other lake users: "Everything is going to come together if they stay offshore farther and stay in deeper water."

As for whether even universal observance of the Wake Responsibly guidelines is achievable, and whether that alone would cool wake boat controversies, Dan Butkus, of Squash Lake, has his doubts. In 2020, at the urging of one wake boat owner, the Squash Lake Association (separate from the lake district) sent a mailer to all property owners suggesting that wake boats stay at least two hundred feet and preferably four hundred feet from shore and avoid repeated passes. "For the most part we're starting to see wake boats stay more toward the middle of the lake," says Butkus. "So people finally woke up a little bit. Do we still have a couple of bad actors? Yes. We can't turn them in because they're not breaking any regulations. Peer pressure? I don't know if that actually works. I have come to the con-clusion that you have to mandate good behavior. There are some who fol-low good practices, but let's face it, a number of people don't pay attention and ruin it for everybody. That's what regulations are for." Forester believes the most probable form of regulation will be local ordinances that take into account specific lake characteristics.

Seeking Harmony

Dave Maturen, Michigan Lakes and Streams Association president, has received numerous complaints about wake boats and has seen their effects on eight-hundred-acre Indian Lake, in the state's southeast region, where he has a home. "The whole aspect is an inharmonious use of everybody's water," he says. "What does that do to the rest of the users? My concern is for users such as fishermen, other boaters, and people in watercraft that aren't motor-propelled. What happens to their quiet enjoyment of that lake? Do you have the right as an individual to go ahead and create something that wreaks havoc with everybody else? The real losers are the recipients of the wave, what's left over after they get done having their fun."

Maturen holds out hope for mutual respect and compromise among lake users, and for the emergence of data on wake boats' true impacts. "Maybe there are places where the depth of the lake, the size of the lake, or the con-figuration of the lake would prohibit or severely restrict the use of wake boats," he says. "I'm hoping the study being done by St. Anthony Falls will produce some data that we can look at. WSIA has their study. We've got some. Now let's take a look at what's appropriate and what's not." He wor-ries not just about the impacts on shorelines and the variety of lake users but also on lake fisheries: "What happens under the water may be more destructive than what happens above the water."

North Lake's Timothy Tyre says his lake district's voluntary boating guide-lines have made a difference—many lake residents welcomed them and most boaters have observed them, notably the weekend no-wake schedule: "There is a small subset that disregards it completely, but the vast major-ity of people on this lake honor it. It has had some impact." In the long run, Tyre believes each lake will need to find its own way to cope with wake boats: "Every lake will have to find a person or a committee to work the issue—people with enough backbone to press ahead and enough credibil-ity to assemble a compelling case for responsible behavior."

One example of a cooperative approach to conflict resolution is Citi-zens for Sharing Lake Minnetonka, a group formed in September 2020. While noting that Lake Minnetonka, southwest of Minneapolis, has a his-tory of supporting diverse water activities, the group's website states that "rapidly growing wakesurfing has had a very significant negative impact on the ability of others to enjoy their form of recreation, as well as [on] lake-shore residents' quality of life."[26] Those impacts include large wakes and loud music played by wake-boaters.

The aim is to give voice to all users so that all can enjoy the lake. The pro-posed solution includes establishing hours when wake-boating is allowed, setting minimum distances from shore for wake boats, and enforcing exist-ing noise ordinances. Specifically, the proposal would forbid wakesurfing before noon on bays measuring 250 acres or less. The group has worked with proponents of motorized and nonmotorized recreation and has engaged with boating industry interests in working toward compromise.

Larry Meddock, meanwhile, asserts wake-boaters' right to use the waters.

The overwhelming majority of the lakes in all of our states are seen as sovereign waters of the states; they belong to the citizens of the states. I believe that everybody has equal access to those lakes. Where you've got a very active lake and lots of people out there, we would encourage dialogue between the user groups to see if we can come up with a compromise. Wakesurfing is going nowhere. It is going to be around a long time. So we need to get everybody together, get everybody on the same page, and see if we can't all get along.

Perhaps the ultimate question is whether wake boats have such outsized impacts—like an amped rock band drowning out a group of a cappella singers—that true harmony can't be achieved. Most likely only time, education, and appropriate regulations will tell.

CHAPTER 10

Stealth Invaders

"In a lot of ways, it's shock and awe." That is how Sandra Swanson describes the reaction to the discovery of zebra mussels in Big McKenzie Lake in 2016. Big McKenzie, in northwestern Wisconsin's Burnett and Washburn counties, is the largest of a chain of three lakes connected by a creek. By 2020, zebra mussels had proliferated in Big McKenzie (1,185 acres) and some had appeared downstream in Middle McKenzie (580 acres). So far, Lower McKenzie (187 acres) has not been invaded.

The McKenzie lakes, known for clear water and good musky and walleye fishing, are popular with visitors from Minneapolis–St. Paul and other nearby urban areas. The first D-shaped, dime-sized zebra mussels were found in October 2016 by a family removing a pier on Big McKenzie. A family friend, a Minnesota DNR staff member involved in boat inspection for invasive species, noticed three mussels on one pier support.

It was a surprise, says Swanson. Since the early 2000s, the McKenzie Lakes Association, of which she is president, had an active invasive species program that included extensive lake monitoring studies, seminars, speakers at association meetings, informational packets, and educational boat landing signage. In 2017, the association received a three-year Early Detection and Response Grant from the Wisconsin DNR to help deal with the mussels. In spring of that year, the association launched a boat inspection and cleaning program at the landings on all three lakes. To measure the mussels' spread, association volunteers installed monitoring plates at the ends of their piers in all three lakes. In Big McKenzie, by September 2017, "people could count maybe thirty or forty zebra mussels on these monitoring plates," Swanson recalls. "Now we've gone to two to three thousand on each of the plates. Our docks and boat lifts are covered with them. Covered. When I pulled my dock out [in 2020], there were hundreds of them on the posts of the dock and on the bottom of the boat lift. The plastic wheels on the very front of my dock had forty to fifty of them."

Thomas Boisvert, former Burnett County aquatic invasive species coordinator and now conservation program manager in Lincoln County, believes the mussel population was likely still expanding as of 2020.

> The population has increased enough that there are several year classes of zebra mussels. So when you look at someone's dock, you're going to see different sizes. The big ones are about three to four years old. The middle-sized ones are about two years old, and the smallest are about one year old. So that tells us the population is showing signs of high recruitment and will likely continue to grow at this time. People have to scrape them off their dock if they want it to look nice. People who want to keep their boats in good condition need to have them on lifts.

He has seen cases where zebra mussels have colonized the water intake on boat motors, causing the cooling system to fail. He has also observed mussels attached to "basically every piece of aquatic vegetation." He has even seen dragonfly nymphs with zebra mussels on them.

By summer 2020, says Swanson, the mussels had migrated to Middle McKenzie Lake: "I was angry when I pulled my monitoring plate out for the first time that summer and saw them. There is a very short creek between Big and Middle McKenzie. We kept thinking, 'It's not going to happen. They're not going to come to us.' We have a lower calcium level than Big McKenzie, but it was not low enough to stop them from forming shells. I was angry. I pulled that monitoring plate out and I wanted to rip it into pieces."

The lake association has formed an aquatic invasive species committee and is attacking the problem, working aggressively to stop the spread to other area lakes, with help from paid boat landing inspectors and numerous volunteers. Still, lake residents are distressed. Swanson observes,

> The people on Big McKenzie are sad. They've almost come to ask, "How do we care anymore?" They're seeing zebra mussels on sticks. They're seeing them on the undersides of rocks along the shoreline. There's a large sandbar coming out from a point, called West Point, that partly divides the

lake. That sandbar is 150 feet long, and it has been a super fun place for people to pull their pontoons up, get out and wade, or sit in the water. There were so many broken zebra mussel shells last summer that they couldn't do that anymore. When the mussels die, their shells break up and form glass-like, sharp surfaces in the sand and on top of the sand. It is frightful.

Some of the greatest threats to our lakes began with cargo ships on the seas of Europe and Asia filling their ballast tanks with water, crossing the ocean to North America, and emptying the ballast water in the St. Lawrence Seaway. That is how two tiny yet troublesome pests came to the Great Lakes: spiny water fleas in 1984 and zebra mussels in 1988. These fast-multiplying creatures spread out through the Great Lakes system and have made their way into hundreds of inland lakes in Michigan, Minnesota, Wisconsin, and other states.

Both species can markedly alter lake food webs, biodiversity, and water clarity. They can interfere with water recreation, and their effects on fisheries are still being researched. Once established in a lake, there is no proven, cost-effective way to remove or control them. That means the only way to limit their effects is to keep them from spreading to more lakes. The next chapter deals with ways to prevent the spread of these and other invasive plants and animals. This chapter introduces spiny water fleas and zebra mussels, explores the effects they have had on the lakes so far, and looks at research on what their impacts might be in the long term.

Spiny Water Flea: Voracious Predator

The spiny water flea is a member of a family of tiny animals, collectively called zooplankton, that float and swim about in lake water. Zooplankton are an important food source for small fish species and for the early life stages of perch, panfish, and larger game fish. The most notable native zooplankton are *Daphnia*, commonly called water fleas, a type of crustacean related to crayfish, shrimp, and crabs. Generally a millimeter or smaller in size, *Daphnia* eat free-floating algae and to a large extent form the base of the lake food web that makes its way up to the large predator fishes: bass, walleyes, northern pike, and muskellunge.

Spiny water fleas, also crustaceans, were discovered in Lake Ontario and by 1987 had colonized all five Great Lakes. They soon spread to inland lakes, mainly by way of the livewells and bilges of fishing boats, on equipment

Spiny water flea, actual size typically about 0.25 to 0.5 inch. (Jake Walsh, University of Wisconsin–Madison Center for Limnology)

such as anchor ropes, landing nets, bait buckets, and scuba gear, and in mud stuck to waders and boots. About one centimeter long, they have a large black eyespot and a long, rigid tail with one to four spines. They are predatory, feeding on a variety of zooplankton species, but especially *Daphnia*.

On inland lakes, spiny water fleas so far have not been the ecological catastrophe that some researchers feared when the species first began to spread. Ben Martin, a food web ecologist and PhD student at the Center for Limnology at the University of Wisconsin–Madison, observes, "I do a lot of outreach, and most of what I end up having to do is calm people's nerves about spiny water flea. As soon as they get into a lake, people panic. But it's a lake-specific issue. For the bulk of ecosystems, they probably won't do much."

Prolific Reproduction

That doesn't mean spiny water fleas are benign. Under the right conditions, their population can explode, because they reproduce rapidly. Females can produce up to ten young every two weeks asexually (without mating). The young are mature and ready to reproduce in about seven days. In autumn, spinies reproduce sexually; the females eject resting eggs that fall to the lake bottom, settle in the sediment, go dormant over the winter, and then hatch when the water warms in spring or early summer. The eggs can live in small amounts of water and can resist drying for several hours. That means they can be spread when a fishing boat from an invaded lake travels to another lake nearby. It is possible for just one female spiny water flea reproducing asexually to establish a new population. At their peak during the warm seasons, in hospitable water, populations have been recorded as dense as 140 per cubic meter.[1]

Spiny water fleas live in deeper, cooler water during the day and at night swim toward the surface to feed. Populations increase from spring into summer and peak in autumn. In warm water when food is available, the males feed while the females reproduce asexually. During this period, all young produced are females. As the water cools and the food supply wanes, more males are generated.[2]

Spiny water fleas thrive mostly in deep, cold, clear lakes, although they have been found in warmer, cloudier waters, such as the highly eutrophic Lake Mendota in the Wisconsin capital of Madison. This suggests that they can survive in lakes with a wide range of physical and chemical characteristics.[3]

Effects on Fisheries

The impacts of spiny water fleas are hard to generalize because lakes and their ecosystems are diverse. One obvious effect is that as they proliferate,

Table 5. Spiny water flea distribution

	Lakes invaded (excludes Great Lakes)
Michigan	2[a]
Minnesota	32[b]
Wisconsin	11[c]

[a] Based on findings reported to Midwest Invasive Species Information Network, accessed October 2020, https://www.misin.msu.edu. May not accurately represent the actual number of lakes infested.

[b] Minnesota Department of Natural Resources, Infested Waters List, accessed October 2020, https://www.dnr.state.mn.us/invasives/ais/infested.html.

[c] Wisconsin Department of Natural Resources, accessed October 2020, https://dnr.wi.gov/lakes/invasives/AISLists.aspx?species=SPINY_WATERFLEA.

they can collect in jelly-like masses on the cables of downrigger balls and the fishing lines that anglers use to troll lures in deep water. A more insidious observation is that they can dramatically change the lower levels of lake food chains. Spiny water fleas, being larger and faster at swimming than *Daphnia*, eat them voraciously and can decimate their numbers. Tom Heinrich, Mille Lacs Area fisheries supervisor for the Minnesota DNR, calls *Daphnia* "the big, dumb, fat, happy cows of the zooplankton world. Spiny water fleas just pick them off like crazy." As the algae-eating *Daphnia* decline, algae can proliferate, reducing water clarity and in some lakes making algal blooms more likely.

The effects higher up the food chain are less clear. Because of their size and the spines on their tails, spiny water fleas are difficult for newly hatched and juvenile native fish to eat. Thus with their usual *Daphnia* prey depleted, the fish have less easily available food; that has potential to impede the fishes' growth. More recent observations suggest that larger fish can eat them readily. For example, Martin has looked to determine which fish eat spinies and to what extent. Early in his research, he dissected yellow perch from Lake Mendota, believed to have among the world's densest spiny water flea populations. He found their stomachs full of spinies. For the next year, he performed food web snapshots essentially every two weeks from May through September and consistently found that yellow perch were eating large numbers of spiny water fleas. He also did feeding trials to actually watch how well fish could consume the creatures. For the first feeding trial, he placed spinies in a glass tank with a yellow perch about four inches long.

"It consumed them with no problem," he reports. "It was a most uneventful feeding trial. I had read about perch spitting these things out multiple times and taking twenty seconds to actually choke the spiny water flea down. In that initial trial, I fed about a dozen spiny water fleas to this fish, and not once did it have any problems. And that was consistent through the rest of my trials. I did about eighty fish—bluegills, yellow perch, largemouth bass—and very few fish in any circumstance ever spit the spiny water flea back out."[4]

This does not mean the presence of spiny water fleas has no effect on fisheries. Next Martin wants to examine connections between spinies and fishes' body condition. Similar to body mass index (BMI) in humans, body condition is essentially a measurement of weight relative to length. Martin is assessing long-term shifts of body condition in different fish species to help determine which have been affected and to what degree. "Body condition is important to reproductive potential," he says. "A big, fat female fish can put out a lot more eggs. If fish are skinnier and not growing as fast, they're also not reproducing as fast, and that can have a trickling effect."

Meanwhile, a study by the Minnesota Aquatic Invasive Species Research Center (MAISRC) at the University of Minnesota, led by assistant professor Gretchen Hansen, explored relationships between spiny water flea and zebra mussels and the growth of yellow perch and walleyes in their first year of life.[5] The study covered nine well-known and much-studied lakes in central and northern Minnesota:

- Winnibigoshshish and Cass, invaded by zebra mussels
- Lake of the Woods, Kabetogama, Rainy, and Vermilion, spiny water flea
- Lake Mille Lacs, both
- Red and Leech as controls (without invasions)

The research team used data from thirty-five years of fisheries studies involving the two fish species to test whether they grew more slowly in the presence of either or both invaders. They found that walleyes on average were smaller (102 mm vs. 116 mm) at the end of their first summer in lakes with spiny water fleas than in lakes without. The growth of yellow perch was not meaningfully different. The results were corrected to account for differences in temperature, which also affects growth rates.

"Smaller first-year size is related to walleye survival and recruitment to later life stages and has important implications for lake food webs and fisheries management," the researchers stated. This is because young fish that grow more slowly are more vulnerable to predators, have lower energy reserves to survive the winter, and take longer to shift to higher-energy sources of food, such as other fish and water-dwelling insects, instead of zooplankton. The minimal effect of spiny water fleas on perch growth suggests the young of that species were better able than walleyes to shift to other food sources as the *Daphnia* population declined, according to the study. The researchers said more work needs to be done across a wider spectrum of lakes to quantify the two invasive species' impact on walleye and perch growth and the exact mechanisms behind it.

Population Dynamics

One research finding of concern is that spiny water fleas may be evolving in ways that make them harder for fish to eat. A study funded by the Great

Lakes Fisheries Commission looked at how the spine might change in response to predation by fish. The researchers found two ways in which the species adapts its spine to ward off predators.

In Lake Michigan, they found that rising water temperature signaled higher risk of predation by young fish; the females then responded by producing offspring with longer spines, outgrowing the size of the fishes' mouths. The study found that this change, called adaptive plasticity, probably helped the spiny water flea successfully invade Lake Michigan. In Canadian lakes, the researchers found that spine length can increase through natural selection. In lakes where most predators are young fish with small mouths, spiny water fleas with longer spines were more likely to survive; that trait was then passed down to new generations. The net effect of both routes of adaptation would be to make spiny water fleas less available as food for young fish. Study researcher Andrea Miehls, of Michigan State, observes, "Spiny water fleas are a great example of why species invasions can be so devastating to ecosystems. We have known for decades that the abundance of invasive species negatively affects food webs; we are now only beginning to appreciate the importance that evolution and adaptive plasticity can play in the negative effects of invasive species."[6]

The dynamics of spiny water flea populations can differ greatly from one lake to another. In Madison's 9,800-acre Lake Mendota, for example, a small population existed for a number of years, but it exploded in 2009. Jake Walsh, a freshwater invasion ecologist now with the University of Minnesota–Twin Cities Department of Fisheries, Wildlife, and Conservation Biology, determined that in previous years the lake's water had been too warm for the species to thrive. But in 2009, one of the coolest summers on record for Madison, the water temperature remained low enough for the spinies to multiply en masse, reaching a density as high as three hundred per cubic meter of water. There was plenty for them to eat, since *Daphnia*, their favored food, were extremely numerous, consuming the abundant algae fostered by phosphorus fed into the lake's watershed. The spiny water fleas feasted on the *Daphnia*, leaving the algae to proliferate. As a result, water clarity in the lake, already quite poor, declined noticeably. The spiny population then persisted and even increased after summer temperatures returned to normal.[7] Walsh observes, "For Madisonians, the spiny water flea represents a threat to how we enjoy our lakes. More algae means more beach closures, stinkier shorelines, sick dogs, and greener water."[8]

Spiny water fleas had vastly different effects on 3,800-acre Trout Lake in northern Wisconsin. There the fleas invaded in 2014 and prospered for a few years, but by summer of 2020 they had essentially disappeared, according to the Center for Limnology's Ben Martin. He notes that in 2007, lake trout stocking had begun to bear fruit, as the fish began to reproduce naturally. The trout population blossomed, and cisco, their favored prey fish, declined. With fewer plankton-eating cisco on hand, *Daphnia* flourished and devoured the algae. "We saw a lot of larger-bodied *Daphnia*, and we gained over a meter of water clarity in that lake," says Martin. Then in 2014, spiny water fleas showed up and fed on the *Daphnia*, allowing the algae to return. "We very quickly saw the water clarity shift back to where it had been for the twenty-five years prior, but we didn't have the same number of cisco," Martin says. "That's because the planktivores were split between the cisco and the spiny water flea."

Meanwhile, with the increase in the lake trout population, the cisco were less abundant; those that remained had less competition for food and so grew faster, to larger than 250 mm (9.8 inches) in length. "That's a big cisco," says Martin. "Lake trout couldn't eat them, and so they were getting to very large sizes, over 400 mm [15.7 inches]. Cisco that size can handle spiny water fleas. So in 2017, we saw the highest abundance of very large cisco we had seen in our forty-year record. We hypothesize that those big ciscos may have eaten spiny water fleas out of house and home." In 2020, in looking at cisco stomach contents, and in towing a special net for catching plankton, Martin found no spiny water fleas at all.

A research team led by Charles Kerfoot, professor of biology at Michigan Technological University, also noted lakes in the northern reaches where spiny water fleas apparently disappeared. They include Fish and Boulder Lakes in Minnesota and Portage Lake, Dead River Impoundment, and Spectacle Lake in Michigan. The researchers emphasized that while gains and losses in spiny water flea territory were happening at the same time, the gains were dominating.[9]

To check the spread of spinies in Wisconsin, in 2020 Martin sampled thirty northern lakes where he suspected the species might live. He found none except for two individuals caught in the plankton net on 1,050-acre Plum Lake. While that caused consternation for Plum Lake property owners, Martin does not believe species will prosper there.

It matters a lot what the base of the food web looks like. In a lot of the northern lakes, the zooplankton populations aren't that big. There aren't that many nutrients coming into the lakes, there is not much algae, and it's a much colder environment, so it's a less productive system. Plum

Lake has very few spiny water fleas right now, and I don't anticipate that to change. There is not enough energy at the base of the food web for them to do that well. At Trout Lake, they had impacts, but Trout is a much bigger lake, and as a drainage lake it receives some nutrients from the surrounding watershed.

Spread by Humans

While some researchers speculate that spiny water fleas are spread by ducks and other waterfowl and birds, little evidence supports that. Multiple studies have shown that humans are the main culprits and that the most important spreaders are boats. Valerie Brady and Donn Branstrator, scientists at the University of Minnesota–Duluth, reported the presence of public boat landings as the largest factor in whether a lake is invaded.[10] Beyond boats themselves, fishing gear has potential to spread spiny water fleas, the researchers observed. They looked to determine how the creatures hitchhike on angling gear. Working on Island Lake Reservoir and Lake Mille Lacs in Minnesota, they counted spiny water fleas that became trapped on fishing lines trolled shallow, on lines trolled deep with downriggers on a steel cable, in a trolled bait bucket and a simulated livewell, and on anchor ropes.

The angling lines together accounted for 87–88 percent of the water fleas ensnared, followed by the downrigger cable.[11] Tests were completed both during daylight when spinies are deeper, and at night when they are closer to the surface. Anchor ropes collected very few specimens because they are stationary instead of being pulled through the water. Towed bait buckets and livewells also collected few specimens because they only skim the surface water.[12]

The Michigan Tech research team found another possible way for spiny water fleas to spread: by way of resting eggs eaten by baitfish and then defecated, still viable, into bait buckets and boat live wells.[13]

Despite all that, while Martin advocates for stopping the spinies' spread, especially to lakes where they would most likely cause disruption, he notes that they might not spread as easily as some believe. Spiny water fleas generally need lakes that are fairly deep and that thermally stratify in summer: warmer, less dense water on the surface, floating on a layer of colder, denser water below. That lower layer gives the fleas the cold-water refuge they need. Even then, not every spiny invasion succeeds. "Just because one spiny water flea makes it to a lake that is suitable, it doesn't mean that individual will get to the deep-water habitat and reproduce," Martin says. "Often, it takes several invasions before they establish."

Zebra Mussels: Crusty Hitchhikers

Zebra mussels are perhaps the most potent stealth missile among invasive species in our lakes. They colonized the Great Lakes, though only minimally in Lake Superior, which is too cold and poor in nutrients to support large populations. From the Great Lakes they have invaded hundreds of lakes, reservoirs, and streams in Michigan, Minnesota, and Wisconsin, and many other states, thriving in waters with high enough calcium content to support development of their shells. These small mollusks with striped shells are easy to detect if attached to a boat, a boat anchor, or other piece of equipment. Much harder to detect are the tiny larval forms, called veligers, that come out of the eggs the adult mussels lay.

Once introduced to a new lake, these veligers, barely visible without magnification, grow up into zebra mussels ready to reproduce in prolific fashion. The Minnesota DNR says a female zebra mussel can produce one

Zebra mussel shells typically measure 0.25 to 1.5 inches. (Paul Skawinski, University of Wisconsin–Stevens Point)

Zebra mussels can attach to and accumulate on almost any hard surface. (Paul Skawinski, University of Wisconsin–Stevens Point)

hundred thousand to half a million eggs in a year.[14] The veligers leave the eggs and float in the water for a few weeks before converting to a juvenile form that settles to the bottom. Juvenile forms on smooth, firm surfaces have the feel of sandpaper. Once attached, they grow into adults; populations can multiply to densities of tens of thousands in a single square yard.

Zebra mussels produce tufts of fibers (byssal threads) that protrude between the shell halves. These threads anchor the mussels to hard surfaces.

Significant Impacts

Zebra mussels are relatively widespread in Minnesota and Michigan lakes; less so in Wisconsin, according to the Center for Limnology, which has judged the majority of the northern Wisconsin lakes either unsuitable or borderline suitable for the species, based on low calcium content in the water.[15]

The mussels' effects on lakes and their ecosystems can be substantial. They form crusts on boat pier supports, boat motors and hulls, buoys, swim rafts and ladders, and any solid object in the water. They can cover the bottom in rocky, gravelly, or sandy areas. They can even establish colonies in muddy bottoms, where hard objects such as rocks or pieces of native mussel shells act as bases on which to attach. From there a colony can spread, shell on shell, to form a carpet-like mat. The shells of dead mussels can wash up and collect on beaches and are sharp enough to cut swimmers' feet. The mussels are found in the greatest densities at depths from six to forty-five feet, but they can be found as deep as one hundred feet.[16]

As dire as that seems, their effects on lake biology can be even more severe. Zebra mussels can encrust the shells of native clams and mussels and even the shells of mobile creatures such as crayfish. They are filter feeders, drawing in water and straining out the algae and some of the small zooplankton on which fish in their early life stages depend. Each zebra mussel can filter a little more than a quart per day; huge numbers at work can ultimately make the water crystal clear. Then, because the sunlight can penetrate farther and with more intensity, underwater plants can grow in deeper areas where otherwise they could not. This can lead to nuisance-scale weed growth and changes to the food web and the character of the lake. The deeper light penetration, along with the nutrients zebra mussels excrete, can foster mats of algae on lake bottoms as well as blooms of toxic blue-green algae.[17]

These effects do not apply to lakes across the board. Nick Phelps, MAISRC director, explains that understanding the impact of zebra mussels is complicated: "Just because you get invasive species like zebra mussels, it's not necessarily going to be the end of the world. I use the analogy of walleyes, not every lake is a good walleye lake, so why would we assume that every lake is going to be good for zebra mussels? . . . Some will just have low-level populations and it's not going to be a big issue, but some will have massive impacts to the ecosystem that they've re-engineered, to

Table 6. Lake suitability for zebra mussels

	Suitable	Borderline	Approx. number of lakes in county
Burnett County	31	59	290
Washburn County	25	42	300

Source: Burnett and Washburn County University of Wisconsin Extension, September 2020. Excludes rivers, springs, and unnamed lakes.

the economy, [and] the management costs to keep it off your boats, off your beaches."[18] Burnett and Washburn counties, next to each other in northwestern Wisconsin, have categorized their lakes according to suitability for zebra mussels to thrive, based on the water's content of calcium, which the mussels need to form their shells.

One Lake's Experience

Minnesota's Mille Lacs Lake is an example of a lake where zebra mussel impacts have been substantial; it also illustrates why conclusions about their effects on lake ecosystems can be difficult to draw. For decades, this 132,500-acre lake in the middle of the state has been a year-round mecca for walleye anglers and all manner of visitors. In 2005, the first zebra mussels moved in; two DNR staff members were scuba diving to document different habitat types in the lake when they found one on top of a boulder. In response, divers launched an expansive search to see whether the mussels had established a population. They found only three more in that year. In 2006, divers searched along survey routes called transects and counted every zebra mussel they spotted. They found specimens at five of twenty locations in the northern part of the lake. After that, zebra mussels expanded rapidly, according to Tom Jones, a DNR regional treaty coordinator for fisheries and wildlife: "By 2008, mussels were found on every transect. By 2009, divers could no longer count all the mussels, so they began counting mussels in one-square-foot quadrats every hundred feet." Density in those places increased by a factor of about thirty every year. In 2012 the average density topped out at 1,269 per square foot; the highest density at any single location sampled was 7,696 per square foot at Three-Mile Reef.[19]

In that same year, researchers also estimated how much of the lake bottom was inhabited by the mussels. They used a camera to check 250 points spread across the entire lake and found mussels at 35 percent of them. "In the southeast part of Mille Lacs, there are a lot of rocky reefs and gravel bars, and most of them had zebra mussels," Jones reports. "In the northwest are mud flats that are clay with four to twelve inches of muck on top; there were no zebra mussels on the mud flats. Between those two areas, going from the northeast to the southwest corner, is what we call the Deep Nowhere. It's all about thirty-five feet deep with a very soft bottom, and there were no zebra mussels there." Around the entire edge of the lake are rocky points, rocky shorelines, and smaller near-shore reefs, all of which had mussels. Since 2012, mussel density has declined by slightly more than half. "It

Zebra Mussel Density
Mille Lacs Lake

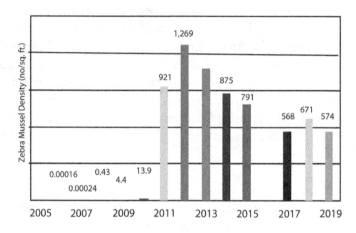

Source: Minnesota Department of Natural Resources. (Eric Roell)

is really common for invasive species to overshoot the lake's carrying capacity and then settle down," says Jones.

Meanwhile, the mussels have changed the way nutrients are processed in the lake. Material that zebra mussels filter is either ingested or expelled as feces or mucus-covered pseudofeces, tending to concentrate nutrients at the lake bottom. "So now there is more filamentous algae," says Jones. "You can definitely see an increase in *Spirogyra,* which is a common slimy green algae." There is also more *Cladophora,* a dark-green branched filamentous algae that grows in shallow water and can wash up and rot in large mats on the shorelines.

Tom Heinrich, of the Minnesota DNR, notes that the pseudofeces are also used by crayfish and many kinds of bottom-dwelling insects, which appear to be prime food sources for smallmouth bass. "One reason smallmouth bass started taking off [in Mille Lacs] is because of the way zebra mussels function," says Heinrich. "The lake's population of yellow perch, the primary forage fish for walleyes, has declined, while smallmouth bass increased. Neither one of those changes is unusual with zebra mussels."

The mussels' effects on the Mille Lacs fishery are unclear; they are suspected (along with spiny water flea) of contributing to changes, but the ecosystem is complex and has been subject to various stresses in recent decades, so that "it's hard to assign cause and effect," Jones says. Water clarity increased noticeably in 1995, suggesting a decline in productivity and probably a decrease in phytoplankton, the lake's primary energy producers. (The increase in clarity has not been attributed to the presence of zebra mussels, according to Jones.) Perch numbers began to drop. Warming water caused a decline in the cisco population, which bottomed out in 2002 and then began to recover. The lake saw strong walleye reproduction in 2007–8, but from 2009 to 2012, low numbers of newly hatched walleyes reached maturity. Walleye fishing was excellent in 2012, and Jones believes anglers likely overharvested fish from the strong 2008 year class; relatively few young fish were there to replace them.

Explains Jones,

> We already had low abundance of forage fish (for walleyes) when zebra mussels and spiny water flea came along. The spiny water fleas especially eat zooplankton, which are food for the small fish. That makes it even harder for little fish to thrive, especially in the face of a bunch of hungry walleyes. And the walleyes are hungrier now. Across this whole time period, they've been getting thinner, and for the last few years they have been as thin as we've ever seen them. All we can say for sure is that these problems started before zebra mussels and spiny water flea. The invasives are probably making things worse, but we can't quantify that.

The mussels also can annoy anglers: A lure or jig that touches the bottom is likely to hook and pull up a clump of three or four zebras.

Scientists are studying Mille Lacs and other lakes to learn more about zebra mussels' impacts on fish. Research led by Gretchen Hansen of the University of Minnesota found that on average walleyes were smaller (101 vs. 116 mm) after their first summer in the presence of zebra mussels. In lakes with both zebra mussels and spiny water fleas, end-of-first-season walleyes measured 88 mm; perch showed no significant decrease in size.[20]

Another study led by Hansen looked at the effects of water clarity and temperature in Mille Lacs Lake on the amount of safe walleye habitat. The study noted that as the lake's clarity has increased in recent decades from improvement in surrounding septic systems, walleye habitat has declined.

That is because walleyes prefer lower light conditions; the lake has large areas of relatively shallow water, and as sunlight penetrated deeper, the fish were left with limited areas of deeper, darker water in which to take refuge.[21]

Lake Carlos, another lake invaded by zebra mussels, is intensively monitored under the Minnesota DNR's Sentinel Lakes project, which has gathered long-term data on twenty-five lakes chosen to represent the diversity of waters across the state. Zebra mussels came in 2009 to this 2,600-acre lake in the west central part of the state. Since then, according to Casey Schoenebeck, who coordinates Sentinel Lakes monitoring, the zooplankton population has declined by 80 percent, but the fish seem mostly unaffected. Schoenebeck noted that walleyes are doing well and that smallmouth bass have increased, as in some other lakes with zebra mussels. Fish do seem to have adjusted their habits, according to Schoenebeck: "For the most part, our growth rates are unchanged, but their location has changed. As the water clarity increases, they're moving deeper. Some of the light-sensitive species, like walleye, are deeper in the lakes now."[22]

Potential Controls

As impossible as it might seem to control an invader like zebra mussels, a variety of remedies are under study. For example, two tests have been conducted to evaluate a biological control product called Zequanox. Proven in laboratory and lake environments, it selectively controls zebra mussels and their cousins, quagga mussels, another Great Lakes invasive species, with minimal collateral harm to other aquatic creatures. The formulation is not a chemical but is made up of dead bacterial cells. When eaten by the mussels, it disrupts their digestive processes and kills them. In 2018, the Tip of the Mitt Watershed Council in Michigan, with funding from the Great Lakes Restoration Initiative, tested the product on zebra mussels on about three acres of an inland lake bottom; it did not kill the mussels but did no harm to other organisms, says Caroline Keson, the council's monitoring program coordinator. However, in December 2020, the Invasive Mussel Collaborative completed a successful test against quagga mussels using different methodology near Lake Michigan's Sleeping Bear Dunes National Lakeshore. The product reduced mussel density by 95 percent. In the succeeding months, the collaborative monitored the site for long-term effects.[23]

Meanwhile, the MAISRC and the Environment and Natural Resources Trust Fund in 2020 began a test of RNA-interference (RNAi) for zebra mussel control. The technique is designed to disrupt the expression of genes important in the spread and establishment of zebra mussel populations. The study aims to reveal genetic weak points in the mussels and help in devising methods of biological control that could be scaled up to open-water applications. The research plan states, "Some lakes that are caught soon after they are infested may respond to chemical controls. But about three to five years after a lake is colonized, dense clusters of mussels become so widely distributed within the lake that the costs and risks to native species of chemical pesticides makes their application impractical. Genetic methods may be effective alternatives in these well-established water bodies."[24]

Another MAISRC study led by Diane Waller, a research fishery biologist with the U.S. Geological Survey, looked at applying inexpensive copper sulfate for zebra mussel control. The researchers in 2019 applied a low dose of copper sulfate to a mussel-infested bay in Minnesota's 14,500-acre Lake Minnetonka. "The study suggests that the treatment effectively reduced zebra mussel veliger density, juvenile zebra mussel recruitment, and live zebra mussel density," according to a MAISRC report. There were some reductions in the density of zooplankton and in the abundance of bottom-dwelling invertebrates, but no adverse effects on native mussels. The bay was to be monitored in 2020 and 2021 to assess the treatment's long-term efficacy and the native creatures' recovery. The research will help determine the lowest amount of copper sulfate that will control the mussels while limiting effects on native species.[25]

As promising as these treatments appear, the consensus remains that the best and most cost-effective way to control zebra mussels, spiny water flea, and other invasive species is to prevent their spread.

CHAPTER 11

Preventing the Spread

Holes in the Safety Net

In summer 2013 a biology teacher kayaking on Rice Lake in northern Wisconsin's Iron County noticed a plant she believed to be curly-leaf pondweed, a nonnative aquatic species. Staff at the county Land and Water Conservation Department confirmed that she was correct. By 2020, the plant had spread widely in the 147-acre lake, growing especially thick in a large bay on the lake's southeast end. Boating in heavily infested areas became nearly impossible. "I tried going through it with my boat," says Robert Kary, president of the Rice Lake Association. "I went about ten yards and the motor got so clogged that I had to pull it up and clear it. There's no way you can get through there in a boat."

It was unclear how the plant got into the lake, as it had no public boat landing. In any case, the mucky bottom of Rice Lake provided ideal habitat for curly-leaf pondweed. It spread rapidly among abundant and diverse native vegetation in the eutrophic (nutrient-rich) water; a plant survey in 2020 had noted lush growth of coontail, fern-leaf pondweed, and elodea. The two dozen lake property owners felt the curly-leaf pondweed had changed the lake's character. They watched year by year as it spread while the association completed the lake management plans the Wisconsin DNR required as a condition of providing an Aquatic Invasive Species Grant to help fund the plant's removal.

Also of concern, according to Richard Thiede, secretary of the Iron County Lakes and Rivers Alliance, were the Turtle River and lakes along its course (the 12,900-acre Turtle-Flambeau Flowage lies several stream miles downstream). Curly-leaf pondweed seeds, along with reproductive structures called turions (winter buds), eventually sprouted in the river and in Pike Lake, the next water body downstream. Residents there acted quickly to remove the plant. Meanwhile the Rice Lake Association tried to control the curly-leaf pondweed by cutting it with a weed harvester, uprooting it by hand, and removing it with a diver-assisted suction system. Ultimately, as the calendar turned to 2021, an application for a substantial grant neared DNR approval.

Lake residents and visitors fear invasive species perhaps more than any other threat to lake ecosystems. Whether plant or animal, they bring change, usually unwelcome, sometimes minor, at other times a major nuisance. To limit their spread, multiple organizations and government agencies have invested heavily in public education, boat and trailer inspections, and boat decontamination initiatives. Despite those efforts, gaping holes remain in the safety net that protects the lakes from unwanted species.

In lake-rich places like northern Wisconsin, Michigan, and Minnesota, prevention of spread is a daunting challenge. The list of nonnative aquatic

Eurasian water milfoil is an invasive plant that can grow thick mats on the water surface. (Paul Skawinski, University of Wisconsin–Stevens Point)

species is long, and in any given year, new threats can emerge. Trailered boats and other watercraft are the most common means of spread, although invaders can also travel on anglers' waders, hunters' boots, and almost any kind of equipment or apparel people use in exploring the waters. Among the three states, there are thousands of named lakes to protect, many with more than one boat landing, some with both public and less-monitored private landings. Adding to the challenge, just one boat carrying a nonnative plant or animal can start an invasion in a lake. Once such a species takes hold, it is usually impossible to remove and, in many cases, difficult and expensive to control.

Invasive species are generally described as those that can take over parts of a lake ecosystem because they are aggressive, spread easily, have no natural enemies, and could harm ecological, economic, or human health. Some were imported years ago and intentionally stocked in the wild. Others were brought to North America in the ballast water of overseas cargo ships and released into the Great Lakes system, from which they spread to inland streams and lakes. Some were sold in the nursery, aquarium, aquaculture, or fish bait trades and were introduced to lakes by accident. Some nonnative plants are quite colorful, pleasing to the eye, easy to grow, and extremely hardy; they likely were imported and sold for those reasons, as ornamentals for gardens or ponds.

Species of Concern

Besides the zebra mussels and spiny water fleas described in the previous chapter, here are some species of most concern to the lakes of the Upper Midwest. The various vegetative species can grow densely, impeding recreation such as boating and swimming. In general, they are inferior to native species in providing food, shelter, or habitat for fish and wildlife.

Eurasian Water Milfoil

This perennial plant is among the most dreaded of nonnative species and the most widespread in the Upper Midwest. It is found in hundreds of lakes; it can be prolific when it takes root in hospitable habitat and sometimes becomes dominant, forming nuisance mats on the surface. The plant has featherlike leaves arranged in groups of four radiating out from the stem; it produces a flower spike with tiny white or pink flowers. It spreads easily because a new plant can grow from a fragment that breaks off and settles to the sediment.

Curly-Leaf Pondweed

This plant has wavy leaves colored from olive green to reddish brown, about half an inch wide and two or three inches long, with serrated edges. It grows a one-inch flower stalk that sticks out above the surface; it reproduces mainly by way of turions. Curly-leaf pondweed generally grows from near-shore shallows out to depths of about fifteen feet. It can thrive in low water clarity and easily invades areas where the lake bottom is disturbed.

Yellow Floating Heart

Bright yellow flowers with five petals appear above the surface on this plant, which has circular or heart-shaped leaves. Seeds are released and dispersed from capsules; plant fragments can start new populations.

European Frog-Bit

This plant, which resembles miniature water lilies, is free-floating but sometimes takes root in shallow water. Its round to heart-shaped leaves with a purple-red underside form a rosette. The flower has three white petals and a yellow center. The plants can form dense mats, often among cattails or other shallow-water vegetation. European frog-bit prefers calcium-rich water with no wave action. It has been found in inland lakes and ponds in southeast and western Michigan.

Starry Stonewort

A form of algae rather than a vascular plant, starry stonewort grows in bushy, bright-green masses in water up to thirty feet deep, anchoring in the bottom with filaments that resemble roots. It is best identified by star-shaped reproductive structures called bulbils, about the size of rice grains, that grow at or below the surface of the bottom sediment. It spreads mainly by way of the bulbils but also from stem fragments.

Purple Loosestrife

This wetland plant is easy to recognize by its spikes of showy purple flowers; it has a four- or six-sided stem and lance-shaped leaves about four inches long. It can grow either in shallow water or on land at the water's edge, and it grows densely, taking over lake shorelines otherwise home to plants with better food and habitat value. Purple loosestrife spreads mainly by way of millions of tiny seeds, easily dispersed by wind, waves, or human activity.

Flowering Rush

Clusters of pink flowers distinguish this plant that, like purple loosestrife, grows densely along shorelines. Dark-green leaves extend up from the roots. The plant spreads mostly by way of underground stems (rhizomes) that grow small, onion-like bulbils from which new roots and shoots emerge. The bulbils can break loose and be spread by water currents; pieces of rhizome can also produce new plants, which can grow to about four feet tall.

Yellow Iris

This plant, known for bright yellow flowers, grows two to three feet tall along shorelines in shallow water. The leaves are broad, flat, and sword-shaped. It spreads by way of rhizomes and seeds.

Hydrilla

While so far not present in the Upper Midwest, this aggressive submerged plant has caused widespread ecological and economic losses in Florida lakes and streams; it is found as far north as northern Indiana. A stocky plant with spiny leaves arranged around a central stalk, it is native to Eurasia, Africa, and Australia and was introduced through the aquarium trade in the mid-twentieth century. It grows rapidly; spreads through fragments, turions, seeds, and other means; and once established is extremely difficult to control.

Rusty Crayfish

Native to the Ohio River basin, these crayfish are believed to have been introduced by anglers who brought them north as bait and at the end of their visit emptied their buckets into the lakes. Identifiable by rust-colored spots on the thorax, rusties do serious damage to water plants; they have been described as "underwater lawnmowers," clipping plants off where the stalk meets the lake bed. They also eat bottom-dwelling insects like mayflies, stone flies, and midges, normally food for young fish. They outcompete native crayfish for food and force them out from their daytime hiding places under rocks, exposing them to predator fish. Rusties also reproduce rapidly—one female can produce nearly six hundred eggs at a time.

Red Swamp Crayfish

These crayfish are identified by their dark-red coloring with bright-red, raised spots; they resemble small lobsters. Native to the Gulf Coast and the Mississippi River drainage basin, they are not yet common in the Upper Midwest, although populations have been found in a few ponds and lakes in southern Wisconsin and Michigan.

Quagga Mussels

Similar to zebra mussels, these mollusks so far are mostly confined to the Great Lakes and surrounding river systems. They can live in a range of water conditions and at great depths. Given their prolific nature, scientists believe that if spread to inland lakes, they could have impacts similar to those of zebra mussels.

Mystery Snails

Chinese, Japanese, and banded species of these golf-ball-sized snails can form large populations and outcompete native lake species for food and habitat. Banded mystery snails prey on fish embryos. The snails' shells can litter shorelines.

Another kind of potential threat is viral hemorrhagic septicemia, a serious disease that has been identified in much of the Great Lakes system and has infected several inland lakes in Michigan and Wisconsin. The virus can cause major fish kills.

This list is by no means exhaustive. State natural resource agencies maintain watch lists of species with potential to expand their range through warming climate, gradual natural spread, or introduction to lakes by trailered boats and other recreational gear.

Voices of Moderation

While these species tend to inspire great anxiety, some scientists caution against overreaction. They contend that the "invasive" label by itself can invite responses to a newly arrived species that do not align with sound natural resource principles. Michelle Nault, lakes and reservoir ecologist with the Wisconsin DNR, observes, "We had historically said that Eurasian water milfoil is going to 'choke' the lake and 'strangle' other plants. It's no doubt that someone hearing this would be fearful and have the mentality that we need to fight against this invader at all costs."[1] And yet, a given non-native plant may create serious problems in some lakes, but not in others, depending on factors like bottom characteristics, nutrient levels, and water chemistry.

Paul Radomski, a research scientist with the Minnesota Department of Natural Resources, states,

As agencies have increased their efforts to educate people not to move species around willy-nilly, we often use language that divides nature into good or bad. And then people interpret that language to mean that species itself is bad, and we should kill it regardless if it's cost-effective or not. The language of war we tend to use gets people motivated to do something. That's where we start losing our rationality; pretty soon pragmatic management goes out the window. You've been calling this organism names for years, and now you've got to go do something about it. I can point to a lot of poor aquatic invasive species efforts that have depleted financial or personnel resources that could have been better devoted to more enduring and meaningful efforts, like shoreland restoration or working on water-quality issues.

A good classic case in Minnesota is curly-leaf pondweed. It has been in the state for over a hundred years. It is what we used to call naturalized. It tends to do better in very eutrophic lakes. For many years we were encouraging the actual planting of it in some of our shallower systems that were heavily influenced by nutrient pollution, as another cover and wildlife food source. Then when we got the aquatic invasive species programs, we started treating it. Probably the worst kinds of treatment were in places where there were no other aquatic plants. In the southern part of the state where we identified curly-leaf stands, people don't like plants in water, and so they went ahead and controlled those populations. They eliminated the only fish habitat they had, and then there were secondary consequences to removing the plants. They might get algal blooms, since now the phosphorus in the water has to go somewhere, and it's picked up by algae instead of vascular plants. Yellow iris in some states has been labeled an invasive species. I know several places in Minnesota where people have taken very aggressive control efforts to remove yellow iris, even though it's really hard to make a case that it's all that invasive.

Scientists also contend that aggressive attempts to control nonnative species risk doing more harm than good. One example is lake-wide herbicide treatments aimed at controlling Eurasian water milfoil. A study in Wisconsin based on data from 173 lakes found that in such treatments, more native species were negatively affected by the herbicide than by the milfoil that was the treatment target. The study authors concluded: "Our comparison reveals an important management tradeoff and encourages careful consideration of how we balance the real and perceived impacts of invasive species and the methods used for their control."[2]

Radomski argues for a case-by-case approach to managing newly introduced species: "You've got a plant here, it doesn't matter its date of arrival or place of origin; it's creating an obvious nuisance condition for our recreational activities. So let's define the problem, define our objective, and define the best way to manage that population. Let's move forward like that, instead of attacking the species just because of the fact that it's new.[3]"

This approach does not minimize the importance of preventing the spread of nonnative species. Says Radomski,

> That should be on the top of our list. That's where you get the most bang for the buck. Many places are doing a great job on that. There are people at public accesses communicating that we should not be moving species around. We're inspecting boats. We're making sure people follow best management practices related to our boating use. Those are important activities, because we are the dominant driver of moving species around our landscape and our waters. If people are educated and change their behavior, we can reduce the risk of species entering places where we would rather not see them. Nobody wants to see zebra mussels, spiny water flea, Eurasian water milfoil, or starry stonewort in their lake. So it's incumbent on all of us to change our behavior to make sure we reduce the movement of species as much as possible.

Pervasive Information

Indeed, education to promote behavior change is everywhere. In Radomski's words, "you would have to be living under a rock" to be unaware of the need to inspect a boat before launching into or leaving a lake. State laws require it, and some local ordinances go further. State natural resource departments, county and local governments, lake associations and lake districts, university extension offices, and others actively promote and advertise Clean Boats, Clean Waters initiatives (Clean In Clean Out in Minnesota). They post websites, distribute brochures and videos, exhibit at fairs and community events, and place news items in newspapers and on TV and radio. They have trained armies of paid and volunteer boat landing inspectors. Prominent signs at landings warn boaters to "Stop Aquatic Hitchhikers" and spell out the requirements and the penalties for failure to comply. "Inspect, Clean, Drain, Dry" is a mantra every boat owner by now should know well.

Evidence suggests that the education campaigns are working well. For example, in Minnesota, DNR inspectors in 2019 found that 96.5 percent of boats arriving at landings had their drain plugs removed, as required by state law; 97 percent of boats and trailers were free of aquatic plants. On the other hand, that leaves small but still meaningful numbers of boats out of compliance. The inspectors also found zebra mussels on 191 incoming watercraft, including twenty-five at water bodies not known to be infested. At roadside check stations where DNR staffers inspect watercraft and related equipment, 19 percent were found to be in violation (a marked improvement from 37 percent in 2012).[4]

Spread by Ballast Water

Wake boats, although less numerous than other types of watercraft, have been identified as a significant risk for the spread of invasive species, notably zebra mussel larvae (veligers). The St. Paul *Pioneer Press* reported that a study by the University of Minnesota's Aquatic Invasive Species Research Center analyzed water samples from 379 boats on two large Minnesota lakes and found that "the ballast tanks of recreational wakeboard boats beat out other leading stowaway suspects, including sterndrive inboard/outboard engines, bilges and livewells, all of which can harbor the larvae."[5]

In addition, the study determined that "zebra mussel larvae that make their way into wakeboard boat ballast tanks take longer to die than those that stow away in smaller areas, such as livewells." The newspaper warned that "it looks like standard minimal tactics for reducing zebra mussel spread—cleaning the exterior of a boat and removing the drain plug— won't be enough for wakeboard boat ballast tanks and some other large-volume water-holding areas. A hot-water flush or five days of drying are the most practical surefire ways to ensure no larvae survive."

On a more optimistic note, three of every four boats inspected in the study carried five or fewer zebra mussel veligers, likely not enough to cause an infestation in a lake. Among boats that carried more than five veligers, the number two culprits after wake boats were large stern-drive engines, followed by traditional outboards, and then the various water-containing compartments. The larger the compartment, the more veligers the researchers found.

State government agencies and boat manufacturers are now evaluating design changes to make it easier for ballast tanks and other water-holding compartments to drain completely and so limit the transport of veligers from one lake to another. Boat industry measures include aggressively promoting a "Clean, Drain, Dry" initiative; hiring an expert to guide the industry on best practices and new boat designs; forming a subcommittee of engineers and specialists to define methods for eliminating species transfer in ballast tanks; and creating a coalition including state agency, boating, and fishing industry members to advocate for federal, state, and regional invasive species legislation.[6]

Meanwhile, a company called Wake Worx offers a Mussel Mast'R inline filtration system that helps keep invasive species from entering wake boat ballast tanks or bags.[7] The company says it can be retrofitted to any boat and has been proven more than 99.7 percent effective.

Meticulous Care

Whether or not boaters know the laws and best practices for preventing the spread of aquatic invasive species (commonly referred to as AIS), the question is whether the vast majority consistently heed the requirements and inspect as thoroughly as they should. Boat inspection means more than pulling off easily visible lake weeds and fragments. There are several less obvious places where weeds can hide. Then there is the matter of

Boat Inspection Points

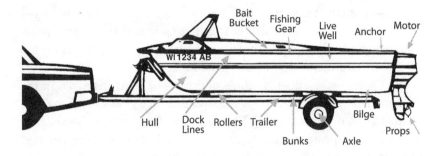

These and other points on boats and trailers should be inspected for signs of invasive species. (Eric Roell)

Table 7. Watercraft inspections for aquatic invasive species

	2014	2015	2016	2017	2018	2019	2020
Minnesota	209,000	330,000	418,000	450,000	470,000	511,000	606,000
Wisconsin	126,548	135,392	148,079	150,645	144,569	149,506	123,798
Michigan	Not available						

Source: Minnesota Department of Natural Resources, Wisconsin Department of Natural Resources.

Note: In Michigan, watercraft inspections are not mandatory and are not performed by state natural resource agencies. Michigan State University Extension and several local natural resource organizations offer voluntary boat cleaning and inspections.

draining water from the motor, bilge, and livewell and checking fishing gear and anchor and mooring lines—especially when leaving a lake known to contain zebra mussels or spiny water fleas. In those cases, the boat should be decontaminated with a bleach solution or a hot-water pressure washer after removal from the water. Otherwise the boat should be allowed to dry for five days before being taken to another lake so that problem organisms will desiccate and die.

This complexity explains why education alone may not suffice; boat inspections at landings provide a critical line of defense in making sure the laws are observed. By way of analogy, "Everybody knows you're supposed to wear your seat belt, and you make a choice: you wear it or you don't," says Adam Doll, watercraft inspection program coordinator with the Minnesota DNR.

The difficulty with inspection is that even though you know what you're supposed to do, just through human error you might miss a spot. You might forget to look under the trailer next to the tire and there's a bunch of milfoil hanging down. Even though you're educated and know you're supposed to do the right thing, you might just make a mistake and miss vegetation or zebra mussels. That's where inspectors really do help. They can serve as a second set of eyes. Inspection programs try to educate boaters on the risks of invasive species. The other aspect is to educate them on how to inspect the equipment, so they're not just looking for the weeds and pulling them off. They're making sure the water is drained out, and maybe sponging the livewell, trying to minimize any movement of material from Point A to Point B.

Essential Inspections

For inspection, a key issue is how thoroughly the landings are staffed during the boating season, which typically runs from May through October. At 183 days, assuming twelve hours of daylight per day, landings ideally would be staffed for 2,196 hours per year. Few lake communities have the resources for that level of protection; even with help from state or county grants to help pay inspectors, many fall severely short. Conversations with AIS coordinators and lake association leaders reveal that some landings are staffed rarely or not at all. Many are staffed only on weekends, holidays, and other high-traffic days, and then often for shifts that do not encompass all daylight hours—anglers are known for arriving at lakes near dawn or just before sunset. Volunteer inspectors may or may not be properly trained; their level of diligence can vary greatly. Staffing tends to be the highest at the most popular, most used, and most sensitive lakes; smaller and more remote lakes typically receive less attention.

Furthermore, Doll observes, "Public boat access is just one of many vectors by which invasive species could be introduced to our waters. Even if we had full coverage at all the public access sites, there are still private accesses, such as at resorts and marinas." Another concern, he says, is the movement of boat lifts and private sales of used docks during the boating season: "People are upgrading their dock; they may sell the old one on Craigslist. Somebody picks it up and puts it in another lake, and it could be covered in zebra mussels. That would never get captured by a watercraft inspection program. We're trying to educate boaters, but we know we can't cover every single vector that might enable invasive species to move into a given body of water."

State-Based Initiatives

In the face of all these challenges, state departments of natural resources make concerted efforts to limit invasive species' spread. Here is a look at the states' boat inspection and other aquatic invasive species programs.

Michigan

Michigan has variety of initiatives to help control the spread of aquatic invasive species and address new species introductions in the early stages. The Clean Boats Clean Waters program began in 2006 as a partnership between Michigan Sea Grant and the state Department of Environmental Quality.

Since inception it has trained 150 volunteers in clean/drain/dry messaging. The program was funded sporadically by way of grants until 2020, when it was reinvented with funding from the Great Lakes Restoration Initiative. It is coordinated by Michigan State University Extension and the Michigan Department of Environment, Great Lakes, and Energy.

A key component is the Mobile Boat Wash program, which provides a trailer-mounted hot-water pressure washer on demand at no charge to lake associations and other groups to help raise awareness of effective clean boating practices. Two staff members operate the unit, talk with boaters, and distribute educational materials. The program has educated more than fourteen thousand boaters through 287 boat-washing events, and 2,196 boats have been cleaned, according to Paige Filice, natural resources educator with MSU Extension.

In 2020 the program added mini-grants of $1,000 to $3,000 for non-profit groups, lake associations, and local governments. They can use the money for actions like producing brochures, banners, signs, and other materials; creating boat wash decontamination stations with plant removal tools; paying staff to inspect watercraft and demonstrate boat-cleaning techniques; and holding outreach events at bait and tackle shops and county fairs. These funds complement the Michigan Invasive Species Grant Program, which offers grants of $25,000 and up to help groups prevent, detect, control, and eradicate aquatic and terrestrial invasive species. Filice is also developing an online library of aquatic invasive species resources, including stories about successful lake association initiatives. Among Michigan's other AIS programs are:

- A requirement for pet shops, nurseries, and other businesses to register and report on nonnative aquatic species they sell. The state's RIPPLE (Reduce Invasive Pet and Plant Escapes) initiative works with these businesses and their customers to prevent the release of unwanted organisms from ponds and aquariums into natural waters.
- The MI Paddle Stewards program trains members of paddling groups and other volunteers to detect invasive species and to clean their watercraft properly.
- Under a DNR Law Enforcement Division program launched in 2019, conservation officers worked 3,184 hours at state-owned boat landings, focusing on boater education and reserving citations for the most blatant offenses.[8]

- An annual Aquatic Invasive Species Awareness Week includes participation in the Great Lakes AIS Landing Blitz, coordinated by the Great Lakes Commission. It covers multiple boat access sites in Michigan, Wisconsin, and Minnesota.
- The Michigan Clean Water Corps Exotic Aquatic Plant Watch trains citizen scientists to detect, monitor, and respond to invasive aquatic plants in lakes. An average of sixty lake groups per year take part.

Providing localized support for invasive species control and prevention are twenty-one Cooperative Invasive Species Management Areas. These are partnerships of groups and individuals that address invasive species impacts on the environment, economy, and human health within a defined region.[9]

Wisconsin

In Wisconsin, the DNR has regional AIS coordinators responsible mainly for education and communication. They support county-based and regional coordinators contracted to the DNR who handle training of watercraft inspectors for their areas. "It's wonderful to have local coordinators because they are familiar with the lake-specific issues, the things people are most concerned about," notes Erin McFarlane, statewide Clean Boats, Clean Waters educator. In a train-the-trainer approach, these coordinators receive refreshers every spring to bring them up-to-date on new inspection forms and materials and the latest areas of emphasis.

Lake associations and other groups can receive Clean Boats, Clean Waters grants, in most cases up to $4,000 per year, to support watercraft inspection programs. DNR lake biologists are available to help groups with grant applications. Gradually, boat inspection has shifted away from volunteers; McFarlane estimates that 70–80 percent of inspectors are now paid through the grants or from other sources: "We rely on the citizens in our state who care about the lakes to fuel our program and keep it going. I'm always amazed at how many groups apply for the grants." For 2021, 153 grants were awarded, totaling $719,000. The DNR also offers Early Detection and Response Grants of up to $15,000 to help lake groups deal with newly discovered aquatic invasive species.

Each spring on the fishing season opening weekend, a Drain Campaign conducted at landings emphasizes the importance of draining boat bilges, motors, and livewells. Each boater receives an ice pack to introduce coolers as a way to keep fish fresh without using the livewell. On the days around the Fourth of July, the year's busiest boating weekend, a Landing Blitz reaches boaters with more comprehensive messages about AIS prevention. Boaters are offered a Stop Aquatic Hitchhikers towel; local newspeople are invited to visit the landings and interview the inspectors. These events are organized by local AIS coordinators.

Other state initiatives include an AIS Early Detector Handbook, and an AIS Snapshot Day on which volunteers inspect multiple sites for invaders. Innovative approaches being explored include detecting invasives by drone surveillance, by testing water samples for species' DNA, and by using specially trained dogs to sniff out unwanted species.

For the past several years the Wisconsin team has conducted a Boater Behavior Study in which boaters and anglers at landings were asked: "Last time you went boating, please describe the steps you took when removing your boat from the water." McFarlane notes that while enough boaters took the survey each year for the data to be valid, there are no clear trends.

However, we can see that draining water from boats and livewells is a prevention step that is taken less often than inspecting or removing. We don't know why that is the case. Perhaps it's because that is a step folks are less aware of, since we emphasized inspecting and removing for years before adding draining. Or maybe folks are less willing to drain because it seems like more work, or they want to leave their fish in the livewell. There is lots of work left to do and data left to collect, but understanding the AIS prevention actions boaters and anglers are taking is an important element of the watercraft inspection program.

Table 8. Prevention steps taken by Wisconsin boaters and anglers

	2014	2015	2016	2018	2019
Inspect	75%	77%	70%	69%	58%
Remove	74%	77%	71%	67%	64%
Drain boat	54%	62%	59%	52%	51%
Drain livewell	36%	52%	46%	43%	49%
Dispose of bait	55%	54%	52%	28%	47%
Took no steps	4%	7%	4%	6%	5%

Source: Wisconsin Department of Natural Resources.

Minnesota

In Minnesota, the DNR delegates inspection authority to local government partners under an arrangement similar to franchising. Partners receive state training and are required to follow state-specified procedures, says Adam Doll. As of 2020, sixty-six of these partner organizations conducted inspection programs, deciding for themselves which lakes to cover and to what extent. Says Doll,

> Some programs are as small as a couple of people staffing one body of water. Others are county-sized, where they make judgments to figure out which bodies of water they want to cover—what months, what days, what hours. June, July, and August are our busiest boating months by far. We want to make sure programs are up and running to cover those months, which coincide with the greatest risk to the lakes, because that is when the vegetation is growing and water temperatures are such that zebra mussels are spawning. Most groups try to start as close to the fishing opener as they can, and usually Labor Day is the end of the main season. Some groups do go farther.

Partner organizations' inspectors have the same authority as DNR staff. By law, a boater approached by an inspector must submit to the inspection or be denied permission to launch. In that event, says Doll, the inspector first calmly instructs the boater about the law and the reasons for inspection. If the person still refuses and launches the boat, the inspector can contact a DNR warden for enforcement.

Beyond the partner groups, the DNR each summer hires forty-six seasonal staff members and an equal number of interns to operate hot-water pressure washer decontamination units at high-profile lakes, mainly heavily used waters known to contain zebra mussels. This helps ensure that boaters leaving a lake do not spread the mussels to the next lake they visit. The DNR also operates roadside decontamination units at rotating sites throughout the summer. "Decontamination is still a challenge because it is largely voluntary," says Doll. "If a decon unit is present on-site, and your boat is covered in zebra mussels, we can require it. But if the boat passes the regular inspection and we don't find anything, then it's up to the boaters, and oftentimes they say no."

Other AIS programs in Minnesota include:

- Permitting for businesses that provide lake services, including training for employees in decontaminating, installing, removing, or renting water-related equipment.
- Permitting and training for commercial bait harvesters who work in waters with aquatic invasive species.
- Dedicating six conservation officers as Water Resources Enforcement Officers, who spend a significant share of their work hours on AIS matters.
- A statewide Aquatic Invasive Species Advisory Committee that reviews and guides species prevention and management work.

Exemplary Efforts

Some counties, lake associations, and other groups go beyond state programs and laws to provide an extra measure of protection against invasions. These initiatives, some backed by regulations and others built on voluntary compliance, exhibit the potential and the limitations of different preventive measures.

Regular Shoreline Inspections

The Two Sisters Lake Property Owners Association in Wisconsin's Oneida County takes a two-pronged approach to AIS control, combining boat inspections at the landing and regular systematic checks for unwanted species along the entire lakeshore.

"Both processes have to work in tandem," observes Joe Steinhage, who has overseen the shoreline inspections for the past eight years. "Fortunately we have not found anything. Yes, we have some rusty crayfish and some mystery snails, but not very many."

On Two Sisters Lake (719 acres, maximum depth 63 feet), the shoreline inspection initiative divides the lake perimeter into twelve roughly equal segments, each assigned to a volunteer. The inspectors look especially for Eurasian water milfoil, curly-leaf pondweed, purple loosestrife, and zebra mussels.

The volunteers inspect twice each summer: during the last week of June or the first week of July, and during the first week of August. They're equipped with a lake map, a sheet of instructions, and laminated pages with color pictures and descriptions of the target species.

Working from a kayak or other watercraft, they inspect out to twenty-five feet from shore or to the ends of the piers, whichever is greater. They collect plant samples by hand or with a rake, looking especially at windrows of plants that have drifted to shore, and checking pier posts, buoys, rafts, boat hulls, and the actual shoreline.

If they find suspect species, they take a sample, place it in a plastic bag for later identification, mark the bag, and note how far from shore and at what depth they took the sample. They report their findings to Steinhage, who shares it with all the volunteers, the association board, and the DNR. "We would love to see other lake associations do the same thing, or better," Steinhage says. "We are very protective of Two Sisters Lake. It is still a pristine water body, and we're trying to do our best to keep it that way."

Full-Time Monitoring at the Landing

A few lake groups have taken full charge of boat inspection at landings, leaving almost nothing to chance. In Wisconsin, for example, the Black Oak Lake Preservation Foundation posts paid boat inspectors at the lake's only landing from 7 a.m. to 8 p.m., seven days a week, from the fishing season opener on the first Saturday in May through October. Black Oak, a 564-acre clear-water lake in Vilas County, just a few miles south of the Michigan border, is home to many multigenerational families with deep connections to the resource, notes Walt Bates, an active foundation member and head of its citizen lake monitoring program.

The lake has some natural protection against new arrivals. "The only public landing is at the extreme southeast corner of the lake," Bates says. "The predominant winds are out of the northwest, and so anything that would come off of a boat at launching or very shortly after, as they motor away, in all likelihood is going to blow right back up onshore in the very near vicinity of the landing. And when you stand at the landing, both to the right and left there is almost a thousand feet of frontage that is nothing but sand. So whatever does wash up anywhere on that shoreline would just lie there and rot in the sun and die." The lake has two shallow, fertile bays, but they already have heavy growth of native milfoil and other plants that would make it unlikely for an invasive plant fragment to fall to the bottom and take root, Bates observes.

Still, the foundation has chosen a rigorous preventive campaign. "We have people on the lake who've had the experience of an AIS-free lake, only to see it get overtaken, to the point where their property values just tanked," Bates says. "And they didn't want to see that again. In fact, one reason they bought on Black Oak was that it is AIS-free, and we hope it will stay that way." About a dozen years ago, the group began staffing the landing nearly full-time with volunteers, but in time they began to burn out. The group's board then looked at the cost to remediate an invasion of a plant such as Eurasian water milfoil and decided that a full-time paid landing inspection program would be a prudent investment.

The program costs about $25,000 a year on top of a $4,000 annual DNR grant. Paid landing staff members are trained in boat inspection techniques and human relations skills by the Vilas County invasive species coordinator. Under a program called Dollars for Hours, every property owner can choose to volunteer ten hours per season as a landing inspector or donate $100; the vast majority opt for the donation.

Inspectors carry a list of area lakes that have any invasive species. "The first question our inspectors ask is, 'Where was this boat last?'" says Bates. "If it came from a totally clean lake, then it receives a very light scan. But if it just came from a lake we know is infested, then we go at it." For example, if the previous lake has zebra mussels or spiny water fleas, the inspectors check the livewell, the bilge, the fishing lines, the anchor rope, the propeller, and any other place where those species might be sequestered. Of most concern are contractors who arrive with boats in spring and fall to install and remove piers and boat lifts. Many come from outside the area; they are under pressure to complete as many projects per day as possible. Bates says, "We have had spectacular saves by our inspectors who have intercepted these people just about to launch trailers that were loaded with weeds."

Partners in Education

Lisa Adams was concerned about the health of three-hundred-acre Big Bass Lake, and especially about controlling the Eurasian water milfoil that had invaded. After retiring from her job as a research scientist in a pharmaceutical company, she devoted spare time to recruiting a few other property owners on the lake, about thirty miles northeast of Ludington, Michigan, to her cause.

She invited Erick Elgin, an aquatic ecologist and water resource educator with Michigan State University Extension, to visit the lake and explain the procedures for an aquatic plant survey. "While we were on the boat," she recalls, "I told him that I was getting frustrated trying to make headway with finding nonchemical approaches to treating invasives in the lake." He advised her to assemble a group representing a number of lakes in the area as a way to gain leverage in seeking grants and promoting policies.

Out of that conversation, in 2019, Irons Area Water Protection Partners was born. Adams reached out to contacts from Loon, Harper, Elbow, Sand, and Cool Lakes and organized an initial meeting to share ideas and discuss mutual concerns. In attendance were Elgin; Chris Riley, a U.S. Forest Service biologist; and Jo A. Latimore, an aquatic ecology and outreach specialist with the MSU Department of Fisheries and Wildlife.

All the lakes except Elbow (a small lake with carry-in canoe/kayak access) had Eurasian water milfoil; Harper Lake also had starry stonewort. "That was an impetus for us to come together," says Adams. "One lake with a public access has starry stonewort. What can the rest of us do to keep it out of our lakes?" At a subsequent meeting, the group decided that an education campaign should be the top priority. With Latimore's help, they secured a $2,500 grant, and they quickly put it to work.

Beside each roadside sign welcoming motorists to Elk and Sauble Township (where the lakes are located), they posted signs reading, "Clean Boats, Clean Waters. Clean, Drain, Dry." At each boat landing, they posted a large sign instructing visitors to clean their boats upon leaving the water. At each landing, they also posted a bright-red alert sign picturing the invasive species found in the lake. Elbow Lake's access received a bright-green sign stating, "No contamination found. Please help us care for this lake by preventing the spread of aquatic invasive species." The group also bought "Clean, Drain, Dry" ads on local restaurant placemats and took a full-page ad in a weekly publication, *Bulletin Board News*, distributed through local businesses.

The Partners group continues to meet and discuss future projects, which could include volunteer boat inspections at the landings. Adams looks at the work to date as a good start and sees the advantage of lake groups working together: "It really is a grassroots thing. When you go to county or township meetings, you can't just be one person from one lake asking for things."

Decontamination Made Mandatory

Burnett and Washburn Counties in northwestern Wisconsin have gone beyond state law with ordinances that require boaters to decontaminate their watercraft if the necessary equipment is available at the landing where they plan to launch. Boaters use a garden sprayer containing a diluted bleach solution that is strong enough to kill zebra mussel veligers but will not harm boat finishes or decals.

The northwest counties have been progressive on invasive species for a number of years. Burnett, Washburn, Polk, and Bayfield counties had ordinances forbidding the transport of such species even before Wisconsin adopted its statewide Invasive Species Rule in 2009, according to Dave Ferris, Burnett County conservationist.

Burnett and Washburn counties upped the ante after zebra mussels were discovered in Big McKenzie Lake in 2016. In the next year, lake associations began placing signage and bleach sprayers at boat landings, but they soon decided that decontamination should be mandatory. "There was a push that came from the bottom up," says Ferris. "The Burnett County Lakes and Rivers Association and individual lake associations came to the natural resources committee of the county board and said, 'We would like to see the do-not-transport ordinance updated to require decon.'" Among those leading the charge was the McKenzie Lakes Association. "We became active very early in the summer of 2017," recalls Sandra Swanson, association president. "We helped write the law. We bought the signs. We bought the tools. We initiated volunteers on all three McKenzie lakes to replenish the bleach every forty-eight hours."

The Burnett County Board adopted the ordinance in 2018, as did the Washburn County Board. Lake associations are responsible for staffing the landings; county personnel can provide some limited support. Large signs at the landings explain the ordinance and spell out the decontamination requirements. Boaters must spray "essentially everything that touches the water," says Tom Boisvert, former Burnett County aquatic invasive species coordinator, now conservation program coordinator in Lincoln County. That includes the bilge, the motor's lower unit, and, ideally, the livewell inside the boat. The county has a hot-water pressure washer to deploy at landings during the summer; in that case a county employee does the spraying.

The program cannot cover all the estimated 175 landings in Burnett County. About twenty-five decontamination units are available (about the same number as in Washburn County) and they are used at about fifty landings per year, says Ferris. The highest-use lakes receive priority. Days and hours of coverage vary from lake to lake; weekends and holidays get the most attention.

Landing inspectors first explain the reasons behind the ordinance. "We try not to make it sound like policing," says Boisvert. "We take an educational approach and help them realize why they should want to do this. You want them to understand the law and why it's important." Boaters who refuse decontamination and launch anyway are reported to county conservation department staff members, who refer the violation to the sheriff's department for enforcement. Typically, the boater first gets a warning by phone call and letter. A violation after that carries a fine and court costs totaling about $250; penalties for subsequent violations increase to as much as $750. Ferris says, "Right now, things are working well."

Consolidating Inspections

A regional approach to watercraft inspections has been demonstrated in one Minnesota county as more cost-effective than inspections based at individual boat landings. The nonprofit Wright County Regional Inspection Program (WRIP) operated during 2018, serving three lakes from a central location in the Annandale area, a popular destination for anglers and other boaters about forty miles west of the Twin Cities.

Chris Hector, then president of the Greater Lake Sylvia Association (GLSA), who became president of the WRIP, characterized invasive species such as zebra mussels, Eurasian water milfoil, and starry stonewort as threats to the area's lake ecosystems, property values, tax base, and tourism economy. The program was a joint effort of the GLSA, the Pleasant Lake Association, the Lake John Association, the City of Annandale, Wright County, the Initiative Foundation, and the University of Minnesota. The premise was that boat landing coverage in the county was not adequate to keep unwanted species from spreading. "If you have ever tried to staff a ramp, you know that other than on weekends it's hard to know when to allocate staff," Hector says. "You're trying to guess when boaters are going to be there." Typically, inspectors would sit idle for long intervals between boat arrivals. Conducting inspections instead at a central location would concentrate traffic and make better use of inspectors' time.

The group secured cooperation from the area cities and townships, the county government, lake associations, and the Minnesota DNR. The program set up shop on a lot in the Annandale Business Park on Highway 55, convenient to anglers bound for the three lakes, each about ten minutes away. Inspectors were on duty from half an hour before sunrise to half an hour after sunset every day from just after the fishing season opener through October. The site was equipped with a hot-water pressure washer for decontamination. Says Hector, "Boaters got an inspection. We put a seal on the boat and gave the owner a tag to put in the vehicle window. Then periodically we drove to the landings and looked for vehicles that didn't have that tag displayed. It was like parking enforcement downtown." Violators were referred to the county sheriff's department or the Annandale police. "Our goal wasn't to find all the people who didn't get inspections," Hector says. "Our goal was to put in boaters' minds that there was a significant chance they might get caught if they bypassed the inspection."

Even while covering the three lakes, the inspectors were idle 82 percent of the time. The lesson was that the site could accommodate many more lakes; the plan for 2019 was to expand to cover eleven lakes. However, the DNR did not authorize the expansion, and the project ended. Hector believes that with some twenty lakes within a ten-mile radius, more inspection sites could have been created. A disadvantage to the approach was that many boaters had to drive more miles to visit an inspection site than if they headed straight to a landing. "At the end of the day, Minnesotans were happy with the status quo," says Hector. "Certainly the DNR was."

Hector notes that lake associations strongly supported the project in concept and financially; many boaters backed it too, but some opposed it vehemently. "My observation is that it was a good thing," Hector says. "We raised a lot of money and a lot of awareness. It was front and center in our county for the three years we were planning and operating it. We were successful in showing the need and in showing that we could do it. I personally think regional inspection is very well suited to the Midwest, where we have lots of lakes to protect."

Decontamination on Wheels

Vilas County in north central Wisconsin's lake region deploys a trailer-mounted hot-water pressure washer decontamination unit staffed with paid interns hired by the University of Wisconsin–Oshkosh. The unit is used on boat landings at four high-traffic lakes, three of which contain spiny water fleas. The county and the university chose the pressure washer because the bleach solution sometimes used for decontamination is not effective against the resting eggs of spinies, says Catherine Higley, county aquatic invasive species coordinator.

At the landings, interns offer decontamination to boaters arriving at and leaving the lakes; boaters are free to accept or decline. Data from interviews with boaters during 2019, provided by Higley, yields a snapshot of their behaviors and attitudes toward invasive species prevention. Of 234 boats encountered, 70 percent were fishing boats and 10 percent were speedboats;

others included kayaks, canoes, personal watercraft, pontoons, and wake boats. Eighty-two boats (35 percent) were decontaminated.

Some boats did not need decontamination because they were being launched for the first time that year or had not been in a lake for weeks or months. About 30 percent of boaters said they had used their watercraft in another water body in the past five days, or planned to do so in the next five days. In eleven cases, decontamination of such boats may have prevented exposure of a lake to spiny water fleas or zebra mussels. On the other hand, because some of these boaters refused to decontaminate, there were sixteen cases in which a lake may have been exposed to those species, Higley reported.

Interviews revealed that 24 percent of boaters were taking decontamination steps on their own: low-pressure washing, chemical cleaning, hot-water pressure washing, wipe-down, and allowing five days for the boat and equipment to dry. All in all, says Higley, "We're doing some good, but we could be doing a little better." Cost is a key barrier to expanding the program: the total cost in 2018, including purchase of the equipment, was $31,835. "There are definitely holes in the system," Higley observes. "If everyone would wait a little while to go to the next lake, that would save a lot of potential exposure, but that's just not the reality. We're doing what we can.

Higley concludes, "We more than doubled the number of boats decontaminated between 2020 and 2021. Perhaps decontamination is becoming more socially acceptable on a local level. So I believe the program has potential to be even more successful."

Permanent Boat Washes

Two state park boat launches on 9,900-acre Higgins Lake in central Lower Michigan have permanent, two-lane boat wash stations available around the clock during the boating season. That's thanks to the Higgins Lake Foundation, a nonprofit organization funded by donations. The group invested a total of $130,000 to build the stations, which include power washers, according to Vicki Springstead, foundation chair.

The first of the wash stations was built in 2009 on state property adjoining North Higgins Lake State Park, the second in 2014 at South Higgins Lake State Park, which has the busiest launch site on the natural glacial lake. Those two launches account for the majority of the lake's boat traffic. Most of the time the stations are not staffed, but large groups of volunteers take part in landing blitzes on and around the July 4 holiday. On July 4, 2020,

375 cars went into North State Park and about five hundred into South State Park, many with boats. In 2018, the foundation and the Higgins Lake Property Owners Association collaborated with the Michigan DNR on a boat-cleaning campaign. "The DNR provided fifty thousand dollars for us to run a program, handing out educational materials and washing boats," recalls Springstead. "The foundation trained the young people we hired as staff and provided them with T-shirts." Workers rotated through Friday, Saturday, and Sunday shifts during the summer.

Stenciling on the roads directs arriving boaters to the wash stations, and signs provide instructions for their use, which is voluntary. "It's an ongoing challenge to try and get boaters to use them when we're not there," says Springstead. "We don't have any type of official enforcement at the state launch sites."

The lake has zebra mussels, Eurasian water milfoil, and, as of 2018, starry stonewort, but the invasives affect only about 2 percent of the lake's area, says Springstead. The foundation owns and operates a suction harvest boat staffed by certified divers. The group publicizes the boat washes and other prevention efforts through its website, a newsletter, radio advertising, and public service announcements.

Strengthening the Net

Despite all efforts at prevention and millions of dollars of annual investment, unwanted species continue spreading. For example, Minnesota's 2019 Invasive Species Annual Report listed eighty lakes or streams newly infested, including twenty-eight with zebra mussels and twenty-six connected to those waters, and eighteen with new growth of Eurasian water milfoil.[10] The gaps in education, boat inspection, and enforcement are clear. The question is how those gaps can be filled, and at what cost, in states with thousands of lakes to protect. Here are a few examples of more stringent measures.

Statewide Inspection

In Colorado, all trailered and motorized watercraft must be professionally inspected by state-certified personnel before being launched in any water after boating in a different state, upon leaving any water in the state known to harbor an invasive species, and at any time a staff member asks to inspect a boat as it is entering or leaving a water body. The state operates seventy-two inspection and decontamination sites.

After inspection, boaters leaving a lake or reservoir receive a Green Seal with a blue or white receipt. White receipts are given at waters with no invasive species; launching at the next lake is then expedited and decontamination is not required. Blue receipts are given at lakes that contain invasives. Boats with blue seals entering waters without invasives are reinspected and in most cases decontaminated—including ballast tanks and inboard or inboard/outboard motors—with a hot-water pressure washer. Upon leaving the lake, boaters must clean/drain/dry the boat, trailer, and equipment; remove all plant matter; and pull all water drain plugs. The inspections and other programs for invasives control are partly funded by Aquatic Nuisance Species Stamps that cost twenty-five dollars per boat for state residents and fifty dollars for nonresidents.

Watchful Technology

About eighty lakes in Wisconsin and Minnesota have installed systems that automatically monitor boat landings around the clock. Produced by Environmental Sentry Protection, the Internet Landing Installed Device Sensor (I-LIDS) systems combine an educational audio recording with an intelligent video camera. Boaters entering a landing see signs that tell them the site is being monitored and explain how they should inspect and clean their watercraft. When they reach the launch area, the boat's presence triggers an audible reminder with messages similar to those on the signs. Meanwhile the video camera captures the bottom of the boat to detect weeds that might be on the boat or trailer; it also monitors whether the boater clears the material away. Once the videos are captured, they are uploaded by way of cellular networks to be stored and accessed remotely. Authorized people then can log onto a website to search videos by attributes such as date, time, and location. Users can quickly review the launch videos to capture landing activity.

Among the largest adopters of the landing surveillance technology is Minnesota's Otter Tail County, which has I-LIDS units at fifteen boat access sites. In 2018, the county aquatic invasive species staff reviewed 33,885 recordings of watercraft activity at those landings. Only five videos were referred to law enforcement for suspected violations, resulting in only one citation. According to the county's 2018 aquatic invasive species program summary, "The small number of suspect violations indicates the public is aware of their responsibility to Clean, Drain and Dispose."[11]

Another prevention technology is waterless cleaning systems for boats produced by CD3. The units combine tethered cleaning tools with software

An I-LIDS boat landing monitoring and information system. (Courtesy of Environmental Sentry Protection LLC)

that reports on when and how often the tools are used. The equipment includes a wet/dry vacuum, a blower, a brush, a grabber, and lights for night operation. Users have the option to scan a QR code or log onto a website to check in on arrival and check out on departure; they also have the choice to sign a social contract to clean their boat in the future and to have the commitment shared on their social media, according to Edgar Rudberg, a partner in the company. Among localities where the devices have been deployed is Hennepin County, within the Minneapolis–St. Paul metro area and home to several popular lakes and more than sixty urban and rural public boat ramps. During a 2017 pilot test at three public access sites, the CD3 systems logged more than six thousand uses of the tools. In the next

year, according to the manufacturer, tool use increased by 50 percent, and
the county found that at landings with systems present, there were 70 percent fewer AIS law violations.

Dave Ferris, of Wisconsin's Burnett County, notes that some lake associations have devised homegrown methods of boat landing monitoring.
Some simply use standard trail cameras, and others deploy more sophisticated security camera systems. One group taking the latter approach is
the Fish Lake Property Owners Association. Steve Johnson, a member of
the association board for the 356-acre lake, devised a monitoring system
of ten hardwired cameras powered by solar panels. Two cameras record
continuously, each monitoring a section of the roadway leading to the landing. Eight cameras simultaneously record when motion is detected. They
are positioned to observe watercraft from different angles, including from
underneath and above, to detect invasive species and see whether the owner
performed decontamination as a county ordinance requires.

The camera array ensures that reliable information about each boat is
collected. "You can have interference at certain times of day from sunlight
shining directly on a camera," says Johnson. "You can have people parking

One type of commercially available watercraft cleaning station. (Ted Rulseh)

in weird spots and blocking out a camera. We got around that by putting up multiple cameras. If I can't see a license plate at one direction, I can see it at another direction. Or I can see what's going on when one person is decontaminating a boat and another person is putting one in or taking one out. We see the entire landing from multiple angles over an area more than a hundred yards long."

The video is continuously streamed wirelessly across the lake to Johnson's cabin, where he can view a composite screen of live and recorded video from all the cameras. "We have the capability to record up to 240 hours of high-definition video per day," says Johnson. "Through our software we can often review an entire day of boater activity video in less than five minutes." If a video reveals an obvious county ordinance violation, a video link can be sent to the county conservation staff for referral to the sheriff's department. However, the emphasis is on education, delivered by signs and by Clean Boats, Clean Waters workers present during limited hours. "The biggest thing is to make sure people don't view this as a Big Brother–type of system," says Johnson. "It's not the government doing this. It's people who care about the lake and who want to make sure we preserve these lakes for generations. We don't want to give tickets. We want to educate and make them want to decontaminate and comply with the AIS law to protect the lake. Our decontamination compliance is now over 90 percent, and it gets better every year."

An advanced technological approach, discussed in Minnesota but not implemented, would fit each boat with a smart tag that communicates by way of a low-strength Bluetooth signal with software in a kiosk at boat landings. At launch, the system would "write" to the boat's tag a date and time stamp and the identity of the lake. In this way, if the boat went to another landing without an adequate dry-off period or without being certified as inspected or decontaminated, an audio message from the kiosk could alert the boater to the necessary steps to take, including a visit to a decontamination site if need be. The net effect would be to monitor landings at all hours and cost-effectively help contain unwanted species.[12]

Eric Lindberg, founder of Environmental Sentry Protection in Maple Grove, Minnesota, which manufactures the I-LIDS technology, argues that the only effective way to keep boats from spreading invasive species is to require them to be inspected or decontaminated, and that automated tools can help. "It's not that there isn't enough money to do prevention," he says. "It takes more vision. It takes ingenuity to implement the technologies needed to do this, and it takes everyone being in agreement that the lakes and the fisheries are important and deserve to be protected."

For lakes without adequate safeguards at landings, he says, "It's not a matter of if, it's a matter of when" unwanted species will come in. "Look at hospital wards that have quarantines. They don't let a family member who comes into contact with a quarantined patient wander around visiting other patients in the hospital. The good news is that most people want to do the right thing. If we implement new policies, people have to understand that we're not trying to inconvenience them. We're trying to protect the resource."

Chris Hector, of the Wright County program, adds, "It's a public policy decision whether we actually think we have an invasive species problem. Without huge advocacy from a government, probably a state government, we're not going to do much more than spot inspections. To be fair, the lakes are a public resource, and you really want to cover a public resource using something that's based on user fees or taxes across all the folks who own the lakes. The reality is, that's not what happens today."

<hr>

I have my own story about the heartbreak—the word is apt—of seeing a treasured lake taken over by a nuisance species. Here on Birch Lake, where I now live, one of the best fishing spots was straight out from the pier of the cottage where my family took vacations. A cabbage-weed bed filled that side of the lake about fifty yards out from the near-shore rock and gravel shallows. In early August, when we paid our week-long visits, the bright-green cabbage was thick enough to be unfishable, even with a weedless spoon. So I would work the deep weed edge, sitting in a boat at dusk, hopping a leech on an eighth-ounce jig.

On that weed edge, when conditions and fortune conspired, I could catch jumbo bluegills, the occasional largemouth bass, and walleyes—once a seven-pounder. From the late 1980s to the late 1990s, a trip to the cottage from our home in southeast Wisconsin was the highlight of summer. Then circumstances kept us from the lake for two years. When we returned, I headed out in the boat on the first evening, looked for the cabbage weeds, and found them—gone. Not trimmed back, not thinned out. Gone. Acres and acres of weeds, every stalk and leaf. Gone.

From my reading, I knew right away what had happened. The culprits were rusty crayfish. I had first seen them while vacationing in 1985 on another northern lake. They darted aside as I snorkeled among the weeds in a shallow bay. When my eye caught a slender white shape on the bottom, I reached down to get it—a walleye skeleton, picked clean, head to tail. One word captures the sensation: chilling. I knew there were rusties in Birch Lake; I would see them beside the pier at night, my flashlight beam revealing the rust-colored spot on each side. For several years there were only a few, but suddenly they overcame whatever had kept them in check. When we arrived after our two-year absence, they had exploded. To explore the shallows with a flashlight after dark was to risk a case of the willies: for each square foot of bottom in the swimming area, a crayfish, or two, or three, prowled along, outsized pinchers held forward.

That wasn't the worst of it. Without the cabbage weeds, it was hard to find the walleyes and the once-abundant perch my son, Todd, and I liked to catch for dinner. That first year back, I poked and probed around the lake like a man wearing a blindfold, the spots I had known now barren, my favorite tactics useless. The nutrients in the water that once fed the cabbage weed had to go somewhere, and they showed up in clumps of filamentous green algae that lay all over the bottom; large, fibrous clouds of it. Now it was impossible to fish a bait on the bottom. If I tried, the leech or worm came back in the vicious pliers-grip of a crayfish, or covered with a musk-scented, greenish-black glob of algae. It was dismal, so much so that despite all the good times the lake had given us, the memories, pictures, stories, and traditions, I wondered: Should we keep coming back? Had rusty crayfish ruined the lake?

The story has a mostly happy ending. As tends to happen in lakes invaded by rusty crayfish, the population explodes until it outgrows the food supply, and then declines. Sooner or later the perch, bluegills, bass, and walleyes learn to eat the crayfish, further reducing their numbers. On Birch Lake, a friends group formed and ran a set of crayfish traps during the summers, taking more crayfish out of circulation. In response to all this, the population hit a kind of equilibrium, higher some years than others, but not appalling. The crayfish became part of the lake food web. Slowly but surely, the weeds began to grow again. Walleyes are abundant. Smallmouth bass are plentiful and grow to trophy size. Perch and panfish, severely depleted after the initial crayfish invasion, are resurgent.

Still, it's hard to forget the trauma of those early years when rusty crayfish dominated. Did they ruin the lake? No. Did they change it? Most definitely. For the worse in some ways, arguably for the better in others, depending on one's preferences. For instance, largemouth bass were plentiful in Birch Lake preinvasion; since then smallmouth bass have prospered, to the delight of many anglers, as the loss of weeds exposed the rocky habitat smallmouths prefer.

As for Rice Lake, which was discussed in this chapter's opening, the association in February 2021 received approval for a three-year, $127,000 DNR grant to pay crews for a project to hand-pull the curly-leaf pondweed. In 2021 the association received an advance payment of $23,700 and removed forty-four tons of the plant in the twenty-four most-infested acres of the lake.

In February 2022, the DNR terminated the grant on the grounds that the Town of Mercer had not provided a suitable public boat access on the lake as required, and that the existing town-owned public access was "not suitable for further development due to the substrate within the riparian area." As a result, the association had to return most of the grant finds already received. The association had hoped a persistent hand-pulling effort could return the lake to conditions that existed several years previously, according to Richard Thiede, of the Iron County association.

While in the long run the introduction of an invasive species is not a disaster by definition, the impacts can be severe and lasting, and the remedies difficult and expensive. Prudence thus favors the scientific consensus to keep nonnative species from spreading indiscriminately. To reinforce that conviction, there's nothing like having watched an invasion alter the character of a lake one knows and loves.

Thiede advocates for more action at the state level to curtail the spread. He observes that states such Maine, Colorado, and Montana vigorously protect their lakes by restricting which boats are allowed access and by carefully inspecting boats from out of state. "We need to do something, and it's not up to the DNR," he says. "It's up to us to demand that our legislators step up and write some realistic and useful protections of our treasures."

CHAPTER 12

Changing Climate,
Changing Lakes

One day in August 2016, residents of the 5,139-acre Lac Courte Oreilles began to notice dead cisco and whitefish floating in the water or washed up along the shoreline. It was the start of a die-off that would last twenty-six days and end with residents reporting hundreds of dead fish of each species. Gulls, eagles, and cranes helped make short work of the dead, but the event did not quickly fade from memory: it became a focus of attention for the Lac Courte Oreilles Lakes Association, the Lac Courte Oreilles Band of Lake Superior Chippewa Indians, and the Wisconsin Department of Natural Resources.

Lac Courte Oreilles (pronounced la-coot-er-RAY), in northwest Wisconsin's Sawyer County, is one of only five lakes in the state that support both cisco and whitefish. These cold-water species are important forage for muskellunge and walleyes and are indicators of lake health. The lakes association tied the die-off in part to unusual amounts of runoff in the lake's watershed, higher-than-normal phosphorus levels in the water, and abnormally high summer water temperatures. In simple terms, the cisco and whitefish died from a combination of too-warm water and too little dissolved oxygen in the relatively narrow band of depths in which they live.

Gary Pulford, vice president of the lakes association and a retired Minnesota Pollution Control Agency regulator, recalls, "People were very concerned. They hadn't ever experienced something like that. It was a massive die-off of both species. There were certain properties where, depending on the prevailing winds, people were spending a lot of time raking up dead fish. Many people had firsthand experience with fish floating up on their shorelines.

"It was a good teaching moment for all concerned, to help folks understand the cold-water ecology of the lake. People started to figure out that cisco in particular are the prey for their favorite fish species. Without the cisco, you could almost kiss the muskies in the lake goodbye, and the cisco also support the walleyes, which are everybody's favorite."

A report on the event from the lakes association and the Lac Courte Oreilles Band states, "The fish kill experienced in the summer of 2016 is an example of the impacts of a changing climate. . . . With continued trends of warmer temperatures and increased severity of storm events, more frequent and more severe fish kills can be expected unless additional measures are taken to protect the lake."[1]

Brett McConnell, an environmental specialist for the tribe who leads a broad and intensive lake monitoring program, said the die-off was all too predictable. "We were out there sampling every single day during the critical period leading up to the event. We saw what was going on with the water chemistry and the oxygen profile. We knew that within twenty-four hours we were going to see fish start floating up to the surface. The science showed it. And it happened."

Researchers expect the changing climate to have diverse impacts on Upper Midwest lakes. Changes are evident today, according to the North American Lake Management Society (NALMS): "For those of us who live, work or play on lakes, we are already seeing the impacts of climate change, from rising water temperatures, loss of winter ice cover, changes in lake stratification, increased evaporation, and more extreme weather events, including droughts and intense rainstorms. Climate change is further complicating already challenging lake protection and restoration efforts, from managing harmful algal blooms to preventing and controlling invasive species."[2] While less dramatic than fish kills, these changes remain significant.

The planet is warming as greenhouse gases, notably carbon dioxide and methane, accumulate in the atmosphere, trapping more of the sun's heat. The United Nations' Intergovernmental Panel on Climate Change asserts, "Scientific evidence for warming of the climate system is unequivocal."[3] The National Aeronautics and Space Administration (NASA) goes further, stating, "The current warming trend is of particular significance because it is unequivocally the result of human activity since the mid-20th century and proceeding at a rate that is unprecedented over millennia."[4]

The National Oceanic and Atmospheric Administration (NOAA) reports that in 2019, the average global temperature was 1.71 degrees F above the twentieth-century average of 57.0 degrees F, making it the second-warmest year on record. The five warmest years on record from 1880 to 2019 have all occurred since 2015, while nine of the ten warmest years have occurred since 2005. From 1900 to 1980, a new temperature record was set on average every 13.5 years; since then a new record has been set every three years.[5] The projected effects of rising temperatures are substantial and widespread. According to NASA:

- The Greenland and Antarctic ice sheets have shrunk; Greenland lost an average of 279 billion tons of ice per year between 1993 and 2019, while Antarctica lost about 148 billion tons.
- Glaciers are retreating almost everywhere around the world in major mountain ranges, including the Rockies, Alps, Himalayas, and Andes.
- Satellite observations show that spring snow cover in the Northern Hemisphere has decreased over the past five decades, and the snow is melting earlier.
- The global sea level rose about eight inches in the last century; in the last two decades, the rate of rise has nearly doubled, and it is accelerating slightly every year.
- The extent and thickness of Arctic sea ice has declined rapidly over the past several decades.
- Since 1950, the number of record high temperature events in the United States has been increasing, while the number of record low temperature events has been decreasing.
- The United States has seen increasing numbers of intense rainfalls.
- Since the mid-eighteenth century, the acidity of surface ocean water has increased by about 30 percent as it absorbed carbon dioxide emitted from human sources.[6]

The lakes of Wisconsin, Michigan, and Minnesota will also see climate-related changes. Globally speaking, the way to mitigate climate change is to reduce emissions of greenhouse gases, largely by using cleaner, less-carbon-intensive sources of energy. And while lake property owners can't do more than that to influence the climate directly, they can take concrete actions that help the lakes resist the changes a warming climate can bring. Michael Meyer, a retired research scientist with the Wisconsin Department of Natural Resources who has a doctorate in wildlife ecology, observes, "The less impact humans have on the lakes, the more resilient those aquatic communities will be in adapting to the new climate." Toward that end, he advocates practices like maintaining and enhancing natural shorelines and limiting nutrient runoff from lakefront properties.

A Look Ahead

Climate change projections are based on Global Climate Models (GCMs), sophisticated mathematical models that simulate the complex interaction between atmosphere, ocean, and land. The models predict the impacts of a range of greenhouse gas concentration scenarios, from low impact to extreme (or "worst case"). Climate likely will not change uniformly across entire regions such as the Upper Midwest; there may be differences from one state to another and even variations within states. However, projections for Minnesota, Wisconsin, and Michigan are broadly similar.

An example of an analysis is provided by the Wisconsin Initiative on Climate Change Impacts (WICCI), a partnership of the University of Wisconsin–Madison Nelson Institute for Environmental Studies and the Wisconsin DNR. Historical trends to date show measurable warming in Wisconsin in recent years. The past two decades (2000–2020) showed the highest statewide average temperatures since well before the turn of the century. Temperatures have increased the most in winter—by 4.11 degrees F from 1895 to 2019—versus 2.07 degrees F on annual average for the same period. All seasons showed increases, the lowest in summer (0.87 degrees F).

Annual precipitation has also increased since 1895: the most recent decade was by far the wettest on record, and 2019 was the wettest year. Precipitation increased in all seasons: by 4.51 inches per year on average, by 1.72 inches in summer, and by 0.57 inches in winter. Along with that, extreme rain events are now more common.[7]

The WICCI projects that even with relatively low greenhouse gas emissions, Wisconsin would warm by 2.5 to 7.5 degrees F in the middle of the century, and by 3 to 9 degrees F in the late century (2081–2100). By mid-century, the number of extremely hot days (highs above 90 degrees F) would triple, and extremely warm nights (never below 70 degrees F) would quadruple.

Annual precipitation (rain and snow) would increase, and by mid-century extreme rain events (four inches or more in a day) would happen once every six to ten years, instead of every ten to fifteen years as in the past. By late century, such events would happen every five years.

Another study, conducted by researchers at the University of Notre Dame over the entire Midwest and Great Lakes region, showed similar results, with strong warming in all seasons, and a strong consensus among models for more precipitation in winter and spring. Summer precipitation showed little systematic change except in late summer, where the projections showed modest drying.[8] These basic seasonal patterns were projected to increase in intensity through the twenty-first century. Worst-case projections show temperature increases double those for the moderate scenario; this illustrates the importance of greenhouse gas mitigation. More intense precipitation in winter and spring, combined with warmer temperatures, would mean substantially more cool-season runoff. It would also mean increased flooding in many Midwest watersheds, although there would be little change in flooding in the northernmost part of the region, where snowmelt runoff is the major cause of annual peak flows.[9]

Diverse Impacts

The projected effects of these changes on lakes are varied and interrelated. Since Upper Midwest lakes are diverse in depth, clarity, nutrient content, dissolved oxygen, fish and plant species, and more, climate change will likely affect some more than others, and in different ways. However, here are some of the impacts of interest to lake scientists.

Less Ice Cover

"The impacts of reduced ice cover are ecologically significant for lakes and their aquatic species," according to Climate Wisconsin, a project of PBS Wisconsin. "Ice cover regulates lake temperatures, dissolved oxygen levels, light penetration, and many other ecological parameters that govern growth and reproduction of species and interspecies relationships."[10] The Environmental Protection Agency asserts, "Reduced ice cover leads to increased evaporation and lower water levels, as well as an increase in water temperature and sunlight penetration. These changes, in turn, can affect plant and animal life cycles and the availability of suitable habitat."[11]

A trend toward fewer days with ice cover is already apparent. The ice on 9,730-acre Lake Mendota in Madison, Wisconsin, has been observed for decades; average ice duration there has decreased by thirty-one days since 1855. The longest ice seasons concentrate in the early years, and the shortest in recent times.[12] Shorter ice cover could reduce the risk of winter fish kills that can occur in shallow lakes when dissolved oxygen in the water is depleted. On the other hand, NALMS says longer ice-free seasons and warmer water mean longer growing seasons for rooted plants and possibly more algae. In addition, "An extended open water season may increase the occurrence and significance of internal [phosphorus] recycling," further promoting growth of algae and plants.[13]

More Algae Blooms

Phosphorus is the main nutrient that drives blooms of algae, notably harmful blue-green algae. Warmer water for more days each year can enhance algae growth, especially if more phosphorus is present. And research shows that more frequent heavy rainfalls would increase phosphorus-rich runoff into the lakes.

Steve Carpenter, director emeritus of the Center for Limnology at the University of Wisconsin–Madison, cites research showing that waterways receive most of their annual phosphorus load from relatively few significant rainfalls each year. He was the lead author of a study that linked major doses of phosphorus "unequivocally" to extreme rain events. The researchers used daily stream discharge records from the U.S. Geological Survey, covering the early 1990s to 2015, to measure phosphorus entering Lake Mendota from two of its main tributaries. Long-term weather data showed that "big rainstorms were followed immediately by big pulses of phosphorus." The period they studied included seven of the eleven largest rainstorms since 1901, a sign that more major rain events already are happening.[14]

Unlike most lakes in northern Michigan, Minnesota, and Wisconsin, Lake Mendota has a large watershed that includes numerous farms, which can be major sources of phosphorus from sediment, manure, and commercial fertilizer. Still, heavy rains in the northern reaches inevitably increase runoff into the lakes, carrying more naturally occurring phosphorus from sediment and leaves, and from pet waste, fertilizer, and other sources. This is true especially where properties are stripped of natural vegetation in favor of lawns, which allow runoff to flow into the lakes essentially unimpeded.

Adding to the problem, research in Europe indicates that warmer water exacerbates the process of eutrophication, in which lakes accumulate nutrients, weeds, and algae as they age. Erik Jeppesen, a professor at Aarhus University in Denmark, observes that algae growth increases because the growing season is longer, nutrients are more readily available, and there are fewer water fleas and other zooplankton to eat the algae. "Cyanobacteria like it hot, which is part of the reason why we're seeing more toxic algae blooms," Jeppesen said. His research suggests that lakes already rich in nutrients are more sensitive to this effect because a warmer climate creates conditions that favor release of phosphorus from the bottom sediments.[15]

On a more optimistic note, a newly released study, led by Grace Wilkinson at the Center for Limnology at the University of Wisconsin–Madison, found that algae blooms are not becoming universally worse with climate change. In studying long-term trends in harmful algae blooms in 323 lakes, the researchers found that slightly more than 10 percent showed substantial intensification in blooms, while 16 percent showed fewer and less severe blooms, and the majority of the lakes showed no major change.[16]

Changing Water Levels

Water levels in Upper Midwest lakes were generally higher than normal in 2020 and 2021, but climate projections indicate that in the long term, those levels will decline as the climate warms, largely because of more evaporation from lake surfaces. Among those studying water levels are researchers with the University of Minnesota Department of Soil, Water, and Climate. They monitored levels on 2,416-acre White Bear Lake in east central Minnesota from July 2014 to February 2017 to see how much water it was losing to evaporation and whether the amount was changing over time. They also looked at historic data and made projections far into the future. Their analysis showed that from 1979 to 2016, evaporation increased by about 3.8 mm (0.15 inch) each year. They project evaporation to increase from 2017 to 2100 by an additional 1.4 mm (0.055 inch) per year, largely because of shorter ice cover. The researchers say their study has implications for evaporation rates and lake levels all across the area.[17]

In another study, Notre Dame researchers examined the effects of climate change on groundwater and surface waters in the Northern Highland region of north central Wisconsin.[18] Alan Hamlet, an associate professor of civil and environmental engineering and earth sciences, reports that the study saw different responses in two basic lake types. In drainage lakes, with a stream flowing in and out, water levels would remain largely stable. In seepage lakes, where the water budget is determined mostly by rainfall/snowfall and evaporation, water levels would likely fluctuate significantly. "We expect lake levels to become more variable, although mostly going down

for seepage lakes, because evaporation is going up," with warmer summers
and shorter duration of ice cover, says Hamlet.

> For Great Lakes water levels, what I take from recent rapid swings in
> water levels is that we should expect to see high variability, where lake
> levels will go up and down dramatically within a few years. Therefore,
> when people ask, "Should we prepare for high lake levels or low lake lev-
> els?" we say, both. If we want to be resilient to what's coming, we need to
> be able to cope with both scenarios, and changes from low to high water
> conditions may happen very quickly. So, for example, if someone chose to
> put in a dock set on concrete pillars and unable to move, that would not
> be prudent, because the lake is going to move. So it would be better to
> have something that's on wheels and can float—something more flexible
> and adaptable to changing lakes levels. In short, plan for variability. Don't
> plan for one direction or another.

Today's high-water levels on Upper Midwest lakes, especially high in
seepage lakes, are related to cyclical changes in the balance between precipitation and evaporation. Carl Watras, a researcher at the Center for Limnology's Trout Lake Research Station at Boulder Junction, notes that lake levels
historically rise and fall when that balance shifts around a tipping point—
the rate at which water drains out of the region as stream and groundwater
flow. The result has been a rise/fall cycle of roughly thirteen years for most
of the past century. Watras and colleagues began studying the lake cycles in
2012, the year when lakes in the north reached record low levels. Their study
initially focused on two seepage lakes in northern Wisconsin's Vilas County
whose high and low levels matched closely. Next they found that groundwater levels have followed a similar pattern, as have the levels in Lake Michigan, Lake Huron, and Lake Superior. The next questions: Why was this
happening across the entire region? And why on a thirteen-year cycle?

To make a complex story simple, they found, with help from atmospheric
scientists at the University of Wisconsin–Madison, a connection to periodic changes in water temperature in the central Pacific Ocean. "That temperature change is like a slow heartbeat, and it affects air flow across North
America," Watras says. "The resulting air flows correlate with the changes
in our lake levels. You've heard of El Niño. That is in the Pacific much farther south, and it has its own periodicity. But there's another system farther
north that seems to be affecting us."

Climate change may or may not be influencing that cycle. "The climate is changing—that's not in question," says Watras. "The question is: How much of an impact will that have on our water levels? Since 1998, we have seen a change in the thirteen-year cycle. It's now closer to twenty years and more extreme. The recent low was a record low, and the recent high is a record high. But it may be decades before we have a definitive answer."

Walleyes on the Wane

Meanwhile, researchers say warming lakes will likely mean shifts in the mix of fish species, most notably at the expense of walleyes. In a Michigan Department of Natural Resources report on the impact of climate change, Kevin Wehrly, a biologist from the Institute for Fisheries Research, observes,

> Our preliminary results suggest that under all future climate scenarios there are clear winners and losers. . . . Warmwater species such as largemouth bass, bluegill and other panfish are the winners and will see an increase in habitat in Michigan. The losers are coolwater species such as walleye and coldwater species such as trout. Fewer stream miles and lakes are expected to support these species under the projected future climate. Our findings match the results found from similar analyses around the country, and our observations are in close agreement with observations from fisheries scientists in Wisconsin and Minnesota.[19]

Andrew Rypel, a fish biologist at the University of California–Davis, notes that the shift from walleyes toward bass is happening already; he observes that largemouth bass fishing tournaments are being held on lakes in Minnesota and Wisconsin that historically have not been prime bass waters. He mentions Minnesota's Mille Lacs Lake, for years considered one of the nation's best walleye fisheries. "It now has this booming largemouth bass population, and it's the site of some big tournaments," he says, calling the lake "a poster child for the future."[20]

A 2016 study by the U.S. Geological Survey and the Wisconsin DNR supports these projections.[21] The researchers related water temperature to suitability for walleye and largemouth bass in more than 2,100 Wisconsin lakes. They noted that in lakes across the state, walleye populations have been declining for the past thirty years while largemouth bass have been increasing. They projected that those trends will continue; walleyes tend to

Table 9. Wisconsin lake suitability for walleyes and largemouth bass

	1989–2014	*2040–64*	*2065–89*
Walleye lakes	184	25	17
Bass lakes	1,236	1,857	1,961
Both species	41	58	59
Neither species	687	208	111

Source: U.S. Geological Survey, accessed December 2020, https://labs.waterdata.usgs.gov/visualizations/climate-change-walleye-bass/index.html.

thrive in cooler, larger lakes, while largemouth bass prefer warmer waters. The researchers made their projections of suitability for the two species using a computer model of water temperatures from 1979 to 2014 on thousands of lakes of different sizes, depths, water clarity, and historical weather. They forecasted future temperatures using climate-change projections.

The study projected that the percentage of lakes likely to support natural walleye reproduction will decline from 10 percent to fewer than 4 percent of Wisconsin lakes by the middle of the century, while lakes favorable to largemouth bass will increase from 60 to 89 percent. On a more optimistic note, "Walleye populations in large lakes appear to be more tolerant of warming than walleye populations in small lakes," noted study author Gretchen Hansen, a research scientist with the University of Minnesota. The study also predicted conditions more suitable for largemouth bass in more than five hundred lakes that at present are not suitable for either species; that means more opportunities for anglers who prefer bass.[22]

More Fish Die-Offs

Warming lakes especially threaten cisco, fish that anglers rarely see but that provide essential food for walleyes, muskellunge, and other predators. Described by fisheries biologists as "swimming sticks of butter" for their high-energy fat content, cisco depend on deep, cold water that is also high in dissolved oxygen.

They live in lakes that in summer thermally stratify, a warm surface layer essentially floating atop a denser cold region that extends to the lake bottom. Because of their different densities, these two layers do not mix. As long as the summer persists, the upper layer is constantly infused with oxygen from the action of wind and waves. Meanwhile, algae and other organic material sink down into the lower layer and decompose, consuming oxygen.

Oxythermal Stress on Cisco

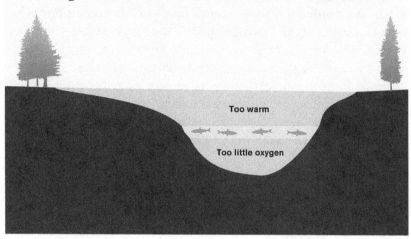

Cisco live in a narrow band of water that is cool and oxygen rich enough for them to thrive. (Eric Roell)

Toward late summer, this deep, cold water can become oxygen poor, and that puts the cisco in what scientists call oxythermal stress.

These lakes are sometimes called two-story fisheries. Fish such as walleyes, bass, and panfish live in the warmer upper layer. Cisco (also called tullibee) live in a band of water at the transition between the warm layer and deeper cold layer; they are essentially sandwiched between water above that is too warm, and water below that is cold enough but low in oxygen. During summer, as oxygen in the lower level declines, the cisco can literally run out of breathing space. John Lyons, a researcher formerly with the Wisconsin DNR and now with the University of Wisconsin, observes, "A lot depends on how long they're trapped down there. If you have a long summer, they're down there longer and there's more time for that finite amount of oxygen to be used up."[23] Meanwhile, warmer summers and continued inflows of phosphorus promote the growth of more algae, ultimately speeding up oxygen depletion in the depths.

The result is a decline in cisco across the Upper Midwest. Minnesota has about 650 lakes with cisco, more than any of the lower forty-eight states, according to the Minnesota DNR. Pete Jacobson, retired fisheries research supervisor, reported, "The statewide decline has been rather steady for the last 30 years. Average statewide abundance, as measured by our survey nets, has declined by about 60 percent over that time period."[24] In Wisconsin, cisco historically were reported from 141 lake systems (individual lakes or lake chains). In a gill-netting survey on 133 lake systems from 2011 to 2014, cisco were captured in only ninety-four (71 percent) of those systems. In the same study lake whitefish, another cold-water species traditionally found in nine lakes, were found in only six. The study suggests that the ranges of both species were shrinking.[25] Meanwhile, Michigan Sea Grant reports that cisco "have declined in, or disappeared from, over 20 percent of their native lakes" in that state.[26]

Taking Action

It is neither cheap nor easy to counteract the effects of climate on the lakes. That hasn't kept governments, university researchers, and others from recommending and taking specific actions.

In Minnesota, DNR scientists are looking to establish refuge lakes for cisco by keeping the land in the watersheds of selected deep, cold lakes in natural forest, thus preventing the increase in phosphorus loadings that would come if the land were converted to homes and cottages. DNR fisheries researchers have found that phosphorous inputs to lakes increase substantially if more than 25 percent of the watershed is farmed or replaced by lawns, roads, and impervious surfaces. They have identified 176 lakes across the northern part of the state that could remain hospitable to cisco as the climate warms in the years to come. These lakes have a virtue beyond fish conservation, according to Carrol Henderson, retired leader of the DNR Nongame Wildlife Program. Fisheries staff members performing regular surveys found large numbers of loons on the lakes in fall; it was later shown that they were feeding on the cisco. "These are important staging lakes for the loons to fatten up a little before the fall migration," Henderson says.

The DNR Forestry Division's Tullibee Lake Watershed Forest Stewardship Project, involving sixty-four refuge lakes, offers incentives to private woodland owners and other conservation partners to keep lands forested. Gary Michael, a DNR private forest management supervisor, notes, "From forest stewardship plans to perpetual easements, there are all kinds of tools out there to help landowners. And if we can keep 75 percent of the watershed as a working forest, then we reduce the threat to tullibee."[27] Landowners who take part receive low-cost forest stewardship plans and can then enroll in the state's Sustainable Forest Incentive program, which pays seven dollars per acre per year to those who set aside at least twenty forest acres for at least eight years.

The DNR estimates it will cost $180 million to protect some three hundred thousand forested acres in the watersheds of all the refuge lakes. Jacobson says, "We have probably a 10- to 20-year window to do this. . . . It's a lot cheaper to protect than it is to restore an ecosystem."[28]

On a larger scale, Denmark researcher Erik Jeppesen emphasizes that reducing nutrient inputs increases lakes' climate-change resilience. Based on the data from European lakes, he and other scientists believe it is essential not just to sustain but to accelerate nutrient reductions, beyond what would be necessary if the climate were not warming. Jeppesen observes, "Even though we humans may change our minds from day to day about whether climate change is real, depending on how hot or cold the weather is outside, the lakes and their biological communities are feeling it. They know global warming is real. . . . It's time to act!"[29]

The Courte Oreilles Lakes Association and the Lac Courte Oreilles Band have stepped up activity to improve their lake's ecosystems ever since the 2016 cisco and whitefish die-off. The first dead fish were spotted by tribal members performing routine oxygen and temperature profiling at six locations on the lake (the east portion of Lac Courte Oreilles lies within the tribal reservation).

One of the first responses to the die-off has been to step up the oxygen and temperature profiling, especially in late summer, to help monitor conditions in the oxythermal layer and detect when the cisco and whitefish might be in jeopardy. "Usually in the middle of July, we go from biweekly profiling to once a week," says Gary Pulford, of the lakes association. "When we start to see the oxythermal habitat shrink, we'll be out there every other day measuring it. We can pinpoint exactly when it disappears. We have had cool summers since 2016, but in each of those summers we have lost all that habitat, at least at our monitoring points." Pulford believes that the fish survive by migrating between the lake's three basins, all more than sixty feet deep, to areas where there is cold water and enough oxygen to sustain them.

To help sustain the lake's cold-water fishes for the long term, the lake association is taking a variety of measures. First is the promotion of natural shoreline habitat and the creation of shoreline buffers to curtail runoff into the lake. Pulford notes that about 40 percent of the 842 properties on Lac Courte Oreilles and the adjoining 221-acre Little Lac Courte Oreilles have inadequate buffers.

The association has also worked with the University of Wisconsin–Stevens Point to inventory farm and forest land in the lake's 125-square-mile watershed. Farms have been inspected, some serious erosion sources have been corrected, and some farmers have voluntarily planted winter cover crops. On the forestry side, the association has encouraged landowners planning to harvest timber to work with companies that follow sustainable practices. The group is also working with three cranberry growers on the lake, asking them to adopt practices that reduce the discharge of phosphorus-containing water from their marshes. One grower has already installed a closed system on one of its marshes, reducing nutrient inputs to an area known as Musky Bay; aquatic plant growth in the bay has declined and water quality has noticeably improved as a result, says Pulford. Meanwhile, in part at the urging of the lakes association and the tribe, the DNR proposed new rules that will establish special dissolved oxygen and temperature requirements for lakes with cold-water two-story fisheries such as Lac Courte Oreilles.

CHAPTER 13

Loons under Stress

In the interest of full disclosure, I will admit that I am grouchy. In the past five days, we have lost three adult loons from our study area. The first of these, having broken his right wing in early July, finally succumbed to that injury, leaving his mate and two almost-grown chicks.

The second loon that we lost was an adult male found in the eastern part of Minocqua Lake. He became incapacitated, beached himself near the Minocqua boat landing, and was picked up in this defenseless condition by Linda and Kevin Grenzer, who took him to Raptor Education Group, Inc. Linda and Kevin could tell from his green droppings and lethargy that he was likely a victim of lead poisoning, and their suspicion was confirmed by a blood test. The folks at REGI started chelation treatment to remove the lead from his blood, but their efforts were too little, too late. A large male weighing about 4.5 kg when he was healthy, he had wasted away to 3.1 kg at his death.

The third death of a male loon occurred just a day later. Chris Rocke, while paddle-boarding on Lake Michigan, ran across a marked loon that stayed very close to shore and seemed reluctant to dive. Over the few days that Chris watched him, this loon became weaker and weaker, to the point where Chris was able to capture him by hand. Chris and his partner, Leva Engel, dropped him off with Linda and Kevin Grenzer, who completed the trip to REGI. It was déjà vu for the Grenzers: another limp, lethargic loon. Like the Minocqua male that had died the day before, this male's lead concentration was so high that it could not be measured precisely. The bird had wasted away from 4.4 kg to 2.8 kg in a period of a week or so, owing to his inability to feed himself. After several seizures (another symptom of lead poisoning), he passed away overnight.

—WALTER PIPER, The Loon Project, September 9, 2020

Loons are Minnesota's state bird. Their images adorn postcards, resort signs, and chamber of commerce brochures. They are immortalized by outsized statues in small lake communities. In gift shops they appear in wood carvings and paintings, on wall decorations and knickknacks. Their long, haunting, three-note wail is the definitive sign by which visitors know they have, at last, arrived Up North. They are without question the symbol and soul of the lake country, an emblem of everything beautiful, pure, and wild. And today in parts of their breeding range, their populations are under stress.

Common loons are widespread across the northeast quadrant of the United States. They are not federally listed as threatened or endangered, and some regional populations are holding steady or increasing. Still, others show worrisome symptoms of decline. In Ontario, in the heart of loon breeding territory, thirty-eight years of data from the Canadian Lakes Loon Survey, managed by Birds Canada and covering more than 1,500 lakes, shows a decline in reproductive success. Researchers have linked the decline to low-pH (acidic) water and associated higher mercury levels, along with stresses related to climate change.[1]

Walter Piper and colleagues in the Loon Project, covering about two hundred lakes in north central Wisconsin's Oneida County, see a substantial increase in chick mortality, lower success in fledging (growing to adulthood), and a decrease in the overall adult loon population.[2] They apply rigorous scientific methods to the study of loons' territorial behavior, reproduction, habitat selection, and population dynamics by following the life histories of loons banded as chicks and as adults.

LoonWatch Wisconsin's latest Loon Population Survey (2020) estimated the statewide population to be 4,115.[3] Michigan, which lists loons as a threatened species, has an estimated three hundred nesting pairs and a total population of just below seven hundred, according to Joanne Williams, state coordinator of the Michigan Loon Preservation Association and Michigan LoonWatch. "They may be slightly declining," says Williams. "We're trying to maintain stability, to help them hold their own." The Minnesota DNR says that state has roughly twelve thousand loons, more than any other state except Alaska.[4]

Challenging Lives

As tranquil as they appear when floating on a lake, intermittently sub-
merging to pursue fish for food, loons face a variety of natural threats. Bald
eagles attack chicks and sometimes adults from above. Swarms of black
flies can beset loons incubating eggs and drive them off the nest. Raccoons,
foxes, and minks steal the eggs; crows and ravens carry eggs off or poke
a hole in the shell and devour the contents. Snapping turtles, muskies,
and northern pike snatch chicks off the water's surface. The Loon Preservation Committee in New Hampshire estimates that about 40 percent of
loon nests monitored in that state over the past four decades have failed;
the most common reason was egg predation.[5]

Unfortunately, the human race that so loves loons can make their lives
and survival all the more difficult. In fact, says Erica LeMoine, LoonWatch
coordinator with Northland College at Ashland in northwestern Wisconsin,
"Human-induced threats are the biggest threats to loons." To one degree
or another, loons are harmed by irresponsible fishing practices, careless
boating, observation at too-close range, deteriorating water quality, unwise
lakeshore development, and destruction of nesting habitat. Loons can be
poisoned and weakened by eating fish containing mercury that originated
with emissions from coal-burning electric power plants. Significant numbers died in their winter territory in the Gulf of Mexico from causes related
to the Deepwater Horizon oil spill of April 2010. Other die-offs have occurred
during fall migration on Lake Michigan after loons fed on round gobies,
an invasive bottom-dwelling fish species contaminated with botulism toxin
related to the decay of *Cladophora* green algae.

Closely Watched

Ecologists refer to loons as a sentinel species—an indicator of high-quality
environments. Lake ecosystems that deteriorate become less hospitable to
loons. For that reason, and because they are so admired by lake residents and
tourists, loons are studied extensively by scientific researchers and observed
by thousands of volunteers.

The Michigan LoonWatch program, started in 1986, employs volunteer
Loon Rangers all over the state. Coordinator Joanne Williams explains,

Some live on the lakes, some just have seasonal cottages, some are snow-
birds. They report to me every year. They watch the loons from the time

they arrive until the time they leave. We count the number of pairs on a lake and how many nest attempts they make. If they lose their first nest, they will often try to nest again. We see how many chicks are alive at two weeks and six weeks and how many fledge to fly off the lake. Our Loon Rangers are very dedicated. They take care of the habitat as well, and educate the people on the lake. We have a database of where the loons are and where the nesting sites are.

Each year that data is shared with the Michigan Natural Features Inventory. A person who plans to build on a lake near a loon nest area, or apply herbicide for aquatic plant control, must acquire a permit and receives information about where loons and endangered species are located.

The Wisconsin LoonWatch Annual Lakes Monitoring Program began in 1978. Its network of Loon Rangers collect long-term data on loon productivity and phenology (seasonal patterns) in the state's northern lakes. The data is shared with the DNR to use as a basis for management strategies. Volunteers attend a spring workshop to learn how to monitor loons and to hear about new loon research. LoonWatch also conducts a one-day Wisconsin Loon Population Survey every five years in July to estimate loon numbers and distribution. A total of 258 lakes are a part of the survey; lakes are randomly chosen based on their size class to represent all lakes that loons might occupy during the breeding season. Volunteers choose from this list of lakes and count the adults and chicks they observe. Results are statistically analyzed. Over time, the survey reveals whether the loon population is stable, increasing, or declining. Since the survey's inception in 1985, the population estimate increased from 2,400 to a high of 4,350 in 2015.[6]

The Minnesota Loon Monitoring Program, in operation since 1994, scientifically analyzes data reported by more than a thousand volunteers from one hundred lakes in each of six index areas. They count all adult loons and chicks during late June and early July. The index areas were chosen to represent different factors that could affect loons and their habitat across the state's breeding range. Those factors include human population growth and density, lake sensitivity to acid rain, the prevalence of roads, and whether the surrounding land is in public or private ownership. After twenty-seven years, the data indicate that the loon population remains stable at about two loons per hundred acres of lake surface statewide. An average of 66 percent of the lakes in the study areas had loons present during the study period.[7]

The Trouble with Lead

The COVID-19 pandemic that began in 2020 in the United States led more people to pursue safe recreation outdoors, and that meant more human interaction with loons, LeMoine observes. Arguably the most insidious and lethal threat is lead poisoning, which occurs after loons ingest lead jigs, sinkers, split shot, and other fishing tackle. As they feed, loons take in pebbles that pass to their gizzard to help grind up and digest the fish they eat. They sometimes pick up lead sinkers or jigs from the lake bottom. At other times they eat fish that have swallowed lead tackle. Loons commonly eat fish up to ten inches long, but they occasionally eat much larger fish. According to the Loon Preservation Committee in New Hampshire, "If these large fish have ingested tackle and are trailing a broken line, they are not able to swim as well as unimpaired fish and are easy prey for loons. The loons then ingest the tackle—which may be a large jig—as well as the fish. Loons may also strike at large fish or tackle as they are being reeled in."[8] Regardless of

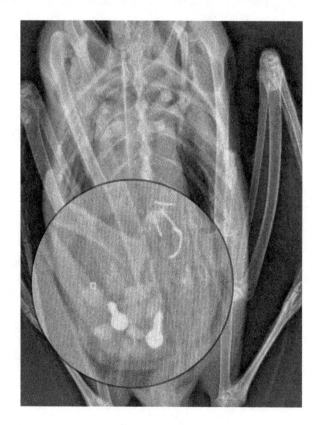

X-ray showing lead fishing tackle that a loon ingested. (Photo by Linda Granzer; X-ray taken at Raptor Education Center)

how the lead is ingested, the effect is fatal. The acid and grinding action in the gizzard dissolves some of the lead, which passes into the bloodstream, organs, and the nervous system.

The Minnesota Department of Natural Resources describes the agonizing process by which the lead causes death. "A loon with lead poisoning behaves strangely. It may fly poorly, have crash landings or stagger onto the ground. The loon begins to gasp, tremble, and its wings droop as lead is carried through its bloodstream. As the poisoning worsens, it eats very little and hides among aquatic vegetation, staying behind when other birds migrate. It becomes emaciated and often dies within two or three weeks after swallowing the lead jig or sinker."[9]

Even the smallest sinker will deliver a lethal dose of lead. Williams observes, "Once the lead gets in the loon's system, the loon is doomed. There is no way out." Most often, a poisoned loon will elude capture until it is so weakened that it becomes largely immobile. By then the lead in its blood has reached a fatal level, and there is little choice but to euthanize it.[10]

This loon treated at the Raptor Education Center in Antigo, Wisconsin, died of lead poisoning from ingesting fishing tackle. (Linda Grenzer)

Scientists have extensively studied lead poisoning in loons. A study by Maine Audubon spanning 1987 to 2012 examined dead loons recovered by volunteers. Poisoning from lead sinkers and lead-head jigs was the largest among thirteen causes, affecting ninety-seven (28 percent) of 352 adults.[11] In New Hampshire, a 2017 study by the Loon Preservation Committee called lead poisoning "the leading cause of mortality in adult common loons" and stated that 48.6 percent of dead adults collected from 1989 to 2012 died by poisoning from lead fishing tackle. Jigs accounted for 52.6 percent of the deaths and sinkers 38.8 percent.[12] The committee concluded that deaths from lead tackle inhibited the recovery of the state's loon population and reduced the overall number of loons by 43 percent: "In other words, instead of the 638 adult loons LPC counted in the state in 2012, New Hampshire would have had an estimated 911 adult loons without mortality from lead tackle."[13]

In Michigan, Tom Cooley, pathologist at the DNR wildlife disease laboratory, examined 376 dead loons over twenty-seven years and found that 16 percent died from lead poisoning, the third-largest cause of loon mortality in the state.[14] Limited research by the Minnesota Pollution Control Agency has linked 11 to 12 percent of loon deaths to lead poisoning, among those with known causes of death.[15]

Carrol Henderson, a board member of the National Loon Center in Crosslake, Minnesota, calls the loss of lead tackle by anglers each year a "cumulative contamination" of lake bottoms. During his career leading the Minnesota Nongame Wildlife Program, staff members received results of necropsies performed on loons found dead. He recalls,

We would have the centers doing the work for us send us bags with the pebbles that were found in the gizzard of each of those loons. In some of those packets there would be two or three dozen little pebbles a little bigger than BB size, and then sometimes there would be a split shot or a lead jig, which would be the reason the loon died. The size of a split shot is almost exactly the same size as those little pebbles. So when loons swallow it, they can't tell the difference. When you take one of those little packets to a sport show and show it to the fishermen, that really makes the point.

As a remedy for lead poisoning, conservation groups advise anglers not to fish when loons are present, as they could become attracted to the bait and swallow it along with a lead weight. Loon advocates also encourage anglers to switch to lead-free sinkers and jigs and ask sporting goods stores to stock those alternatives. The DNRs in Minnesota and Wisconsin sponsor "Get the Lead Out" initiatives that educate anglers about the hazards of lead, advocate nontoxic alternatives, and list vendors of those products.

Anglers can choose a variety of nontoxic jigs, sinkers, and other items in different metals, alloys, and other materials. Tungsten, a highly popular choice for ice-fishing jigs, is denser and harder than lead and helps anglers feel the bite more effectively. Tin, bismuth, and steel alloys are lighter than lead and so require a larger split shot or jig for the same purpose; on the other hand, some anglers report having adjusted to fishing with less weight for a more natural bait presentation. There are even sinkers made from natural stone, and biodegradable weights made of a lead-free compound that if lost underwater slowly degrades and disappears. Fishing tackle containing zinc is not recommended because zinc is also toxic to wildlife.

Recent years have seen attempts to ban or limit lead fishing tackle. Those initiatives on a national scale have not succeeded, but six states—Maine, New Hampshire, Vermont, New York, Massachusetts, and Washington—have banned or restricted the use or sale of lead items. Lead restrictions face opposition from some industry groups. For example, the American Sportfishing Association has opposed bans on lead tackle on the grounds that there is not enough data to justify them, nontoxic tackle items cost more and do not perform as well, education can encourage anglers to use lead responsibly or choose nontoxic items voluntarily, and only a small number of loons and other water birds die from lead poisoning.[16]

Walter Piper, leader of the Loon Project and a professor of biology at Chapman University in Orange, California, offers a contrary argument.

If lead poisoning were a freak occurrence, like a lightning strike, we could justifiably shrug and move on. . . . But a clear pattern has developed in northern Wisconsin: many adult loons die each year from acute lead poisoning. . . . Loons can survive ingestion of fishing tackle. In fact, their powerful digestive systems have been shown to grind up steel hooks and swivels. Lead is different. Lead kills loons, eagles and other wildlife that swallow it, because it has rapid, powerful effects on the brain and nervous system and cannot be quickly broken down or expelled by the body. I think it is time to see lead poisoning as less of a freak occurrence and more of a regular—and probably important—source of loon mortality.[17]

Piper notes how lead poisoning is most detrimental to adult loons.

The adults are the ones churning out the young. If you kill adults, that hurts the population even more than killing chicks. Whatever chicks that adult would have produced, it's not going to produce. The lake can't produce any more chicks until that adult is replaced. If all of us who love loons can't stop using lead sinkers, then maybe we don't love loons as much as we said we did. We're at a place where we can't spare any adults. So if there are simple things we can do to keep from losing our loons, let's do them.

Hazards of Monofilament

Lead isn't the only threat loons face from anglers. Though less common than lead poisoning, entanglement with monofilament fishing line can kill loons. Joanne Williams calls line entanglement "the most prevalent reason for loon rescues" in Michigan. If a loon eats a fish with line still attached, the hook can get stuck in the skin or mouth, or make its way to the esophagus or stomach, according to the Adirondack Center for Loon Education: "When it feels a line dangling from its mouth, a loon will fling its head around or scratch at its head trying to free itself from the line. This can make things worse, as the line then wraps around the bird's beak, head, leg, or wings."[19] Without immediate attention, says Piper, a hooked loon is likely to die: "It could die from entanglement. It could die from some sort of infection. It could be unable to feed itself."

The Loon Project shares one story of entanglement:

Mike Henrichs and his three grandkids were out fishing on Birch Lake when they noticed an adult loon behaving strangely and found that it was dragging a bobber and fishing line. . . . They could see that the loon was not diving to avoid them, as a loon normally would, so they approached the bird, grabbed hold of the fishing line attached to its leg, and pulled it onto their boat. They then patiently cut the line off the loon's leg, around which it had become tightly wrapped, dislodged a hook from the loon's bill, and joyously released it. . . . The Henrichs were unaware of this loon's current status and, naturally enough, presumed he was a resident of Birch Lake. In fact, the entangled loon . . . was a father with a chick and territory [on a nearby lake] to defend. His sudden entanglement and consequent inability to return to his breeding lake put his mate in a bind. She became the sole provider and defender of their seven-week-old chick.[18]

Conservationists recommend that anglers stop fishing, or move to another place, if a loon comes close, since in its curiosity the loon might follow and strike at a lure. Anglers are also advised not to throw scrap fishing line into the water, to pick up any abandoned fishing line they see, and to report any loon that has become entangled to a wildlife rehabilitation center.

They are also admonished not to feed loons by tossing them minnows or small fish, says LeMoine, of Wisconsin's LoonWatch. If fed in this way, loons become habituated to people and will approach them and beg: "When they see a human, they will beeline toward that human and think they're going to get a free meal. That's how they get tangled in monofilament line and swallow lead tackle. You can't feed a loon and not have that happen. A fed loon is a dead loon."

Conflicts with Boats

Irresponsible boating is another threat to loons. Carrol Henderson, of the National Loon Center, notes that the "two most traumatic weekends for loons" are Memorial Day, when they are nesting and curious people approach too close, and the Fourth of July, "when everyone is out tearing around in their boats, waterskiing, Jet Skiing, or using wake boats."

On a busy Saturday in July, Minnesota Department of Natural Resources conservation officer Jacob Swedberg was patrolling Big Floyd Lake near Detroit Lakes when he spotted a loon. As he watched, a jet ski turned sharply toward it, then chased the bird as it raced across the lake, trying to get up enough speed to take off. "The loon was just about out of the water when the jet ski finally turned away from it," Swedberg said. "So, of course, I go flying over there and was like, 'What the heck are you doing?'"[20]

Loons build nests on shore next to the water, on islands, or above the water on mounds, beaver lodges, or artificial nesting platforms. They are ungainly on land and essentially shuffle on their breast to climb onto the nest to incubate their eggs. Wakes caused by watercraft can wash eggs off nests. Another danger is that powerboats can strike loons swimming on or just below the surface. Chicks are the most vulnerable because they do not dive as deep as adults: "They pop up like little corks," says LeMoine, "and therefore they can't get away from a fast-moving boat." Personal watercraft can be especially problematic if driven carelessly because they move at high speed and can easily enter shallow water where loons often hunt for food.

Even quiet kayak and canoe paddlers can pose a threat. LeMoine says, "During nesting season, loons' first line of defense is to remain inconspicuous. They'll take what we call a hangover position on the nest; they try to get really low and close to the water. If a canoe or kayak comes close to the nest and startles them, they will leave their nest suddenly, and they could knock the eggs off with them. Boom, the eggs are in the water, and the loon has no way to put them back on the nest. That is a huge loss." Even if the eggs remain on the nest, the loon will not come back until it feels safe. In the meantime, the eggs are vulnerable to nest predators like ravens and crows and exposed to the elements; in some weather conditions, the eggs may simply become unviable.

Most boaters do not intentionally chase or harass loons; more often they simply drive too fast and fail to notice loons nearby. Experts advise boaters to stay at least two hundred feet away from loons and to avoid making large wakes near them, or along shorelines. Loons let people know when they are getting too close. One typical response is the penguin dance. "They'll sort of run across the water on their feet," says LeMoine. "They're actually expending a lot of energy to do this. They're not doing it for our entertainment. They're doing it because they're scared."[21] Another sign of distress is a soft tremolo call that sounds like a quavering laugh; it is a signal to stay away.

Threats to Habitat

As more lakefront forest land is built up, natural habitat is lost, in some cases at loons' expense. Piper, of the Loon Project, observes that some nesting habitat tends to remain on even heavily developed lakes, since the marshy, boggy areas that loons prefer are either unappealing or completely unsuitable as building sites. Still, the loss of natural shoreline can affect loons in less direct yet meaningful ways.

Notes Henderson,

Shoreland management is really important. Too many people buy a lakeshore lot and then rip out the trees and wildflowers and plant bluegrass down to the water's edge. They're letting all the fertilizer from the lawn wash into the lake, and they're not protecting the lake from erosion. The other thing that happens when people clean up a shoreline is they see emergent vegetation in the water as weeds. They illegally cut down the bulrushes and cattails and whatever they have in front of their property so they have a nice clear view of the lake. So they eliminate the shoreline habitat for fish that possibly are spawning there, and that's also the edge where loons normally place their nests.

Overdevelopment also tends to degrade water clarity, as shoreline erosion adds silt to the water, the inflow of nutrients promotes algae growth and algal blooms, and heavy boat traffic stirs up sediment from the bottom. "Loons are visual predators," says LeMoine. "They need to be able to see their prey to pursue it. Loons used to be found in southern Wisconsin and northern Illinois, Iowa, Indiana, and Ohio. They're not there anymore because those lakes became so developed and the water quality so poor that there was no place for them to nest, and they weren't able to locate fish." Development gradually pushed the loons farther north.

In addition, more lakeshore homes have increased the prevalence and changed the behavior of certain predators of loon nests. Raccoons, crows, and gulls are attracted by refuse people leave outside; some then ultimately find their way to loon nests.

Climate Change

The changing climate could eventually force loons out of some or all of their Upper Midwest breeding territory, according to a 2019 report from the Audubon Society.[22] The region comprises the southern part of loons' summer home. A warmer climate would tend to force the territory boundary northward into Canada, which even today is where the most loons nest. The Audubon report projects that by 2080, if the climate were to warm by an average of 3 degrees C (5.4 degrees F) above preindustrial levels, loons would lose 56 percent of their summer range and 75 percent of their winter range.

While that projection is based on computer models of climate that contain substantial uncertainties, other climate-related changes are more immediate. "To me," says Henderson, "the most dramatic impact would be potential changes in water clarity." Henderson points to what is known in Minnesota as the "carp line," running roughly northwest from the southern Twin Cities to the state's western border. South of that imaginary line, the lakes tend to be shallower, warmer, less clear, and less suitable for loons. "A problem we face if the water gets warmer is that it would enable rough fish like carp to spread farther north than they are currently found," says Henderson. "Once you get that turbid water created by the carp, it's kind of a losing battle. We might see decreased loon productivity moving farther north with any projected changes in climate."

LeMoine wonders whether some detrimental effects are already appearing. For example, leading climate change scenarios predict an increase in heavy rain events of two inches or more. "We have had extreme rain events happening pretty much every year," she says. "A big storm dumps a bunch of rain on the lakes, and it drowns the loon nests." Climate scenarios also predict more intense times of drought. That could drop water levels enough at times to expose land bridges to small islands where loons nest, allowing access by land-dwelling predators, says LeMoine.

A climate change report from Wisconsin's Focus on Energy stresses the importance of protecting loon habitat throughout the state's breeding range. Projected lower water levels in seepage lakes (those that depend on rain and groundwater) have potential to alter the quality of nest habitat, the report says: "Adaptation strategies to reduce potential negative consequences of a changing climate should include preserving existing critical nest habitat by managing shoreline development and habitat loss."[23]

Signs of Trouble, Reasons for Optimism

Evidence and viewpoints differ on the future of loons on Upper Midwest lakes and elsewhere. Piper and his colleagues on the Loon Project rely on direct observation of banded loons in a relatively compact area. The findings give them cause for concern, while data from elsewhere in the United States offers room for optimism.

During twenty-seven years of investigation and statistical analysis of data, Piper and colleagues have observed an 11 percent decline in the weight of loon chicks and alarming increases in chick mortality: 31 percent among chicks younger than five weeks, and 82 percent among chicks five weeks and older. Fledging success has declined by 26 percent. While there has been no noticeable change in the numbers of breeding pairs on territories in lakes, the population of nonbreeding adult loons (called "floaters" because they travel from lake to lake) has dropped by 46 percent. This largely accounts for a 22 percent overall decline in the adult loon population.

At present, Piper can only speculate on the root causes of these declines. "We know that reproductive success is suffering, so definitely the fisheries are one suspect," he says. The lakes in his area are fished heavily, and he would like to investigate whether loon reproduction is falling more substantially on lakes with fewer small fish. "The fact that the chicks are smaller over time, and that there are fewer two-chick broods, suggests that they're not getting as much food as they need; they're dying of starvation or some weakness related to starvation." Another possibility is that adult loons are spending more time than in the past protecting chicks and steering them away from human disturbances, at the expense of time spent feeding their young. However, Piper freely admits, "The real answer is that we don't know."

In his more optimistic moods, Piper looks to the success of loons in parts of the country outside the Midwest. He sees evidence that loons are thriving in Maine, New Hampshire, Vermont, and Massachusetts, where there have been double-digit increases in adult populations in the past decade. While in Minnesota and Michigan the populations are "merely stable," he says, "things look so good for the species in New England that, even after considering the slightly negative recent trend from the Upper Midwest, we must conclude that overall the U.S. loon population is doing fairly well."

Unfortunately, it's not possible for the growing New England loon population to help resupply the Midwest. That is because young loons do not disperse far from the lakes where they were raised. "The stability of the Upper Midwest loon population relies solely upon the successful reproduction of Upper Midwest adults," Piper says. "In other words, we are on our own. Still, the mere fact that loons are reproducing well and expanding their population somewhere is heartening. It suggests that factors causing the decline in the loon population in Wisconsin might be local ones, not sweeping ones, like climate change. Or it might mean that factors that could lead to loon population declines, whatever those factors are, can be reversed by intense local conservation efforts."[24]

In the end, the preservation and prosperity of loons depend on responsible human behaviors and the protection of natural lake environments. LeMoine observes, "If our aquatic habitats are degrading, loons probably are not going to be on those lakes. They are a marker of healthy aquatic ecosystems. Taking care of them is taking care of us."

CHAPTER 14

Pressure Rising

A few years ago, our neighbors moved to assisted living and their kids decided to rent the house on our lake to vacationers. Our side of the lake is zoned recreational. In our county that means an owner can rent out a house by the week and, indeed, we have new neighbors every week. In large homes with numerous beds, it can become quite a crowd. We often count over a dozen cars on-site; it seems each car contains multiple passengers, and often a dog.

Turnover day is Saturday, and new vacationers arrive with enthusiasm and excitement. It's a new party every Saturday night that often lasts until the wee hours of Sunday. Our phone starts ringing by early Sunday with other neighbors asking about the invasion and artillery of the previous night. The bigger the group, the louder and longer the party.

Sunday morning can sometimes be quiet as they sleep it off, but the real fun starts in the late morning. The rental pontoons and Jet Skis arrive then, and by early afternoon it looks like Dick's Sporting Goods threw up all over the front lawn. By later in the afternoon, someone discovers it's five o'clock somewhere, and the flames of the previous night's raucous event are rekindled.

By Wednesday, the evidence of the success of Party Week begins to pile up in our trash container, since they often fill theirs by the beginning of the week. Recycling is a foreign concept for many, and the growing volume of beer containers is truly impressive. Time is short and life on the lake must be lived to the fullest.

Despite warnings and lake maps, lakes like ours that have rock bars often extract some compensation from inexperienced boaters, and our tow rope is always kept handy. If everyone is lucky, the encounters with shallows are made at slow speeds, so no one is tossed into the drink.

As the week draws to a close, the drive to finish the vacation on a high point becomes intense. This is usually when they bring out the fireworks, ignoring the rental agreement terms that forbid them. The end of the pier is nearly always the launch point, but wind direction is seldom factored in, and the launchers' expertise is always minimal. Neighboring boat and boat lift canopies are nicely ventilated by the falling embers, and boat carpeting takes on an interesting speckled look with the random burn spots.

Thankfully, by Friday night, some adults come to their senses and begin to pack up the evidence of the week of too much fun and misdemeanor. The crowd generally departs in a rush on Saturday morning, and for a couple precious hours our shoreline is blissfully quiet. Then the next wave of energy arrives in the late afternoon to begin the revelry anew.

—Neighbor of a short-term rental house on a lake
in Vilas County, Wisconsin

Pressure on lakes continues to rise, in part through longtime trends, and in part from causes few could have envisioned ten or even five years ago. Lakes have been developing steadily for decades; every year fewer vacant shoreline lots remain, and large year-round homes are replacing seasonal cottages. In recent years, as mom-and-pop resorts with housekeeping cabins have been closed and sold off, more property owners rent out their homes for a week or two at a time through services like Airbnb and VRBO (Vacation Rentals by Owner). Perhaps most significant, the COVID-19 pandemic created a boom in home sales in lake-rich counties and elsewhere. People looked to escape to the safety of rural areas, especially the lake country, and soon found they could work remotely from a Northwoods getaway just as well as from a home in the city or the suburbs. The result of these trends: greater demand and stress on the water resources and ecosystems.

Conflict over Short-Term Rentals

As demand rises for homes in the lake country, short-term vacation rentals are proliferating. The online rental industry has grown substantially in the past decade or so. Many owners of lake homes and cabins have turned them into part-time or, in the summer season, even full-time lodging establishments. Some enterprising people, and some commercial enterprises, buy lake properties as investments just for that purpose. Others rent their places only occasionally to help pay property taxes and generally lower the cost of owning a lakeside getaway. Some have converted garages to sleeping quarters or, on larger lots and where zoning allows, have added a guest cabin separate from the home. These rentals sprang up so fast that they outstripped most local governments' ability to regulate them for health, safety, and environmental protection. Neighbors often complain of "party houses" occupied by a dozen or more visitors instead of only an average-sized family. Rental groups come and go, some quieter, more courteous, and more respectful to the lake than others. Complaints about trash, trespassing, parking issues, and noise are common.

Granicus, a company that helps local governments improve services and engage effectively with citizens, sums up the ways in which short-term rentals affect communities. On the one hand, they can stimulate local economies through the money visitors spend in stores, bars, restaurants, and other businesses. Often attractively priced, they can draw more tourists to areas that otherwise might not be highly attractive destinations, or that face a shortage of hotel and motel rooms. On the other hand, these accommodations may compete unfairly with traditional lodgings because the owners may or may not have to pay staff, are less regulated, and in some communities do not pay room taxes.[1] In addition, even in communities that regulate and tax short-term rentals, many "fly under the radar" unless reported to local authorities.

Apart from neighborhood disruption, lake residents and local officials worry about effects on the environment from transient visitors who have no substantial connection to the lakes and no stake in their protection. A key issue is the impact on septic systems. The Rosi & Gardner law firm in

Table 10. Short-term rental properties in a sampling of counties

Minnesota		Wisconsin		Michigan	
Cass	352	Vilas	1,247	Antrim	778
Otter Tail	179	Oneida	478	Charlevoix	1,267
Crow Wing	575	Sawyer	447	Cheboygan	347
Wright	138	Washburn	206	Gogebic	277

Source: Granicus, March 2020.

Traverse City, Michigan, notes that some short-term rental owners, in order to accommodate more people and so reap more revenue, add sleeping quarters in "bonus rooms" and on sofa beds. Because septic systems are sized to the number of bedrooms in the home, the worry is that parties of renters far beyond the homes' normal occupancy could overload those systems.[2]

Ulrik Binzer, general manager of compliance services with Granicus, echoes the concern. "The reality is that if you had lake cabins used sporadically by families over the summer, and now they turn into units that are rented to large groups every day over the whole summer, you have a lot more people contributing to the septic system. If the septic system seeps into the lake, that can have an environmental impact."

Increasingly, communities that permit short-term rentals impose septic system requirements. Typically, they mandate a system inspection before a permit is granted. "They inspect the system to ensure that it is actually working," says Binzer. "Then, based on the system capacity, they will set a maximum occupancy level for the house. If the owner then starts advertising a maximum occupancy that's higher than what is allowed under the permit, that would be a code violation." As a condition of permit renewal, communities may require the owner to prove that the septic tank is being pumped out at an appropriate interval. Jeffrey Goodman, a planning consultant with Granicus, sees some communities taking more technical approaches to septic system regulation, considering for example that short-term rental guests tend to use septic systems more intensively than resident families.

Then there is the issue of greater stress on the lake itself. "A lot of people who rent lake houses are there to go boating," Binzer observes. "If everyone brings a speedboat, suddenly you have multiple people driving up and down the lake. Obviously there are environmental impacts." Those can include effects on fish and wildlife, shoreline erosion from boat wakes, and nuisance noise. An additional concern is the spread of invasive species brought by outsiders' boats; a remedy for that is to require renters' boats to be inspected before launch.

Other stresses from short-term rentals may not affect the lake itself but can affect the character of lakefront neighborhoods or communities. One is impact on housing affordability: when homes and cabins become investment properties, prices tend to rise, potentially shutting some traditional home buyers out of the market. The same is true on the rental side: owners seeking more profit may convert long-term rentals to short-term, reducing the supply of long-term lodgings and driving up their rental rates.

Many counties and townships face pressure from permanent and seasonal residents to place strict controls on short-term rentals. Local officials generally hesitate to make the rules too stringent; they aim to balance neighbors' concerns against owners' rights to earn income from their properties if they so choose. Common features of short-term rental ordinances include:

- Licensing, with fees to cover mandatory property inspections before a license is granted
- Requirements to collect room and sales taxes
- Proof of suitable commercial liability insurance coverage
- Required safety equipment such as smoke and carbon monoxide detectors
- Local contact person easily available to deal with neighborhood complaints
- Maximum number of nights a rented home can be used in a given year
- Maximum and minimum lengths of stay
- Parking regulations
- Sealing and storage of trash containers to prevent nuisances and wildlife intrusion
- Establishment and enforcement of noise ordinances or quiet hours
- Authority to revoke a permit after a specified number of substantiated complaints

In the big picture, the Granicus representatives suggest that communities adopt a strategic vision in dealing with short-term rentals, asking questions such as: What is the optimum ratio of primary residents to vacationers? What are the minimum expectations for short-term rental hosts in terms of property maintenance and respect for neighborhoods? How much boating traffic and other usage can a given lake sustain? Says Goodman, "Having that kind of vision can help local communities make good choices."

Booming Home Sales

Growing interest in the lake country does not end with vacationers. Starting in 2020, COVID-19 and the lowest mortgage interest rates in years combined to create extreme demand for homes and property in northern counties, especially on lakefronts. Real estate agents in Michigan, Wisconsin,

The COVID-19 pandemic helped create a boom in home sales in the northern lake country and elsewhere. (Ted Rulseh)

and Minnesota reported low inventories of properties for sale; from mid-2020 on, buyers jumped on listed properties and often paid a premium for them.

Sandra Ebben, manager of the First Weber Realtors office in Rhinelander, Wisconsin, observes, "We are experiencing a lack of inventory unlike anything I have seen in my forty years of selling real estate. When Wisconsin shut down in March 2020 because of COVID-19, we expected the market to slow down. Things were quiet for a couple of weeks, and then we saw an unexpected increase in buyers looking to purchase homes, cabins, and land in the Northwoods. The story was the same for most of these out-of-area buyers: they wanted to escape the virus and the protests in their cities. They also realized that they could do their jobs remotely and felt the Northwoods was a safer place to live. New property listings were getting multiple offers as soon as they hit the market; buyers were making offers sight unseen and offering over list price."

The demand continued through autumn and winter of 2020, ahead of the spring season when people traditionally list their properties for sale. For example, Minnesota saw sales increase by nearly 16 percent in November

2020 over the previous year, according to Chris Galler, CEO of Minnesota Realtors. People able to work from home especially sought out the lake country. "The mid-counties of the state, which would include Detroit Lakes, Fergus Falls, Alexandria, Brainerd, Baxter—those areas that have lakes and woods and a lot of outdoor opportunities—have done very, very well during the pandemic," said Galler. "People have decided that's a lifestyle that they like and they think they'll be able to continue."[3]

That experience was widely shared among Michigan real estate professionals and their counterparts in Minnesota and Wisconsin. At first they worried that social distancing would limit showings and that people who had lost their jobs would sell their second homes. Instead, the boom market took hold.

In Wisconsin, Krystal Westfahl, director of the Minocqua Chamber of Commerce and Visitor's Bureau, says many new home buyers are longtime summer tourists who used to stay at resorts or campgrounds. She saw a shift in values among those relocating to the lake country: "A lot of it has to do with getting back to their family, grounding themselves in nature. I really do think that people are becoming more grounded about who they are, what they need, and what their necessities are in life. A lot of that can be found here, and that is propelling people to move north."[4]

At the same time, builders in some areas struggled to meet demand for new homes in the north: "Some homebuilders in resort areas of Michigan, such as Emmet, Charlevoix, Cheboygan, Otsego and Grand Traverse [counties], are booked for two years ahead, and they are not talking with new clients to create projects," according to Janet Chambers, executive officer at the Home Builders Association of Northern Michigan.[5] Some real estate agents suspected that the home construction backlog was impeding the sale of vacant lots: people hesitate to invest in property if they might have to wait years to build.

Boating Explosion

With growing attraction to the northland, all kinds of outdoor equipment sold briskly during the pandemic as people looked for social-distanced recreation. Canoes, kayaks, paddleboards, and bicycles saw increased sales. So did all-terrain vehicles and motorized watercraft.

The National Marine Manufacturers Association (NMMA) reported that powerboat sales in the United States reached a thirteen-year high in 2020. More than 310,000 new boats were sold, up 12 percent over 2019 and the most the industry had sold since before the Great Recession of 2008. Frank Hugelmeyer, NMMA president, observed, "2020 was an extraordinary year for new powerboat sales as more Americans took to the water to escape pandemic stress and enjoy the outdoors safely. For the first time in more than a decade, we saw an increase in first-time boat buyers, who helped spur growth of versatile, smaller boats—less than 26 feet—that are often towed to local waterways and provide a variety of boating experiences, from fishing to water sports."[6] The 2020 sales included:

- An 8 percent increase in personal watercraft: 82,000 sold. Personal water-craft sold at relatively affordable prices are often considered a gateway to boat ownership.
- A 20 percent increase in wake boats for wakesurfing, wakeboarding, and skiing—13,000 sold.
- A 12 percent increase in freshwater fishing boats and pontoons boats—143,000 sold. These boats accounted for about half of new powerboat sales.

The NMMA expected boat sales to remain at historic highs in 2021 as manufacturers filled a backlog of orders from the previous year. More fishing boats, those most often trailered from lake to lake, heighten the need for education and enforcement to prevent the spread of invasive species. More boat traffic in general creates greater potential for shoreline erosion and other disruptions.

In an incident reported in Minnesota, Brian Hovland, a water patrol officer on an Otter Tail County lake, estimated he saw about one hundred boats anchored on a sandbar, with a few hundred people playing in the water, listening to music, and visiting. "The only way you could truly count it probably would be with a drone, because it's a pretty big area," he said.

"For me, I see it as a safety issue, just because if something were to happen, you can't get in there. There's just too many people. It's dangerous even to go in there to try to enforce stuff, let alone if there was an accident." The crowding has led to an increase in complaints about noise, trespassing on private beaches, and boats blocking access to residents' docks.[7]

Sandra Swanson, president of the McKenzie Lakes Association in northwest Wisconsin's Burnett County, noticed the impact on her three-lake chain. Water made clearer by a zebra mussel invasion brought denser weeds that grew toward the surface earlier in summer than in the past. "COVID-19 drove people from Minnesota and Illinois to come to their cabins in an effort to isolate," Swanson says. Many arrived in March, April, and May. Docks and boat lifts were in the lakes earlier; more and larger boats were on the water. Wake boats and personal watercraft "cut the weeds off, pulled them out by their roots, so we had more dead vegetation around all three lakes than we've ever seen before," reports Swanson. "The boats were out earlier in the day and later in the evening. People were asking, 'Why do I have all this dead vegetation on my shoreline?'" Visitors at homes rented through the online services also rented watercraft, used them heavily, and, unfamiliar with boating rules, often passed too close to docks and shore-lines. Lake association members went door to door delivering copies of boat courtesy and safety rules.

Michael Engleson, executive director of Wisconsin Lakes, calls recre-ational boating conflicts "by far the number one issue that I hear from our members right now." There is conflict between motor-boaters and users of canoes, kayaks, and stand-up paddleboards. With traffic and bigger boats, in the face of high lake water levels, shoreline erosion has become a larger concern. "Wake boats are the most obvious problem, but people have begun to realize that maybe it's even more than wake boats," says Engleson. "Cer-tainly it's the waves that are causing the most consternation."

The issue is by no means unique to Wisconsin. For example, Minnesota Department of Natural Resources conservation officers in 2020 reported "a marked increase in the number of people recreating on the water," along with an increase in water-related complaints, boating rule violations, and on-water emergencies.[8]

Stress on Fisheries

Meanwhile, there is evidence that fishing pressure is leading to overharvest of some species, notably walleyes and panfish, those most favored as table fare. For example, a 2019 study by researchers at the Center for Limnology at the University of Wisconsin–Madison cited a "hidden overharvest" as contributing to a decline in walleye populations in many Wisconsin lakes. The researchers found that "40 percent of walleye populations are over-harvested, which is ten times higher than the estimates fisheries managers currently use," according to graduate student and lead author Holly Embke. She says this is largely because "for the last 30 years, resource managers have focused on fish abundance and not fishery productivity when calculat-ing harvest limits."[9]

The harvest includes hook-and-line angling by some "1 million recreational anglers who account for about 90 percent of the total annual harvest on the state's 900 'walleye lakes.'" The remainder comes from Native American tribal members who spear walleye on about 175 lakes each spring, exercising treaty rights to hunt and fish within what is known as the ceded territory, essentially the northern third of Wisconsin.

The researchers say their results point to a need for better assessment, and regulation of recreational fisheries in a time when the environment is changing, with a warming climate, more development on lakes, and loss of habitat.

"Nature has changed," says Steve Carpenter, director emeritus of the Center for Limnology. "The climate now is different from what it was in the 1980s and it's not going back. That means habitat is decreasing and, on average, walleye stocks can't take the harvest levels they have seen." On a positive note, he observes that Embke's way of estimating walleye production can help point the way to a more secure future for an iconic fishery.

Walleye harvest is also a concern in Minnesota, where as of September 2021 a bill pending in the state legislature would direct the DNR commissioner to reduce the statewide daily bag from six fish to four. Although the DNR Section of Fisheries did not initiate the proposal, the DNR supported it and representatives testified in favor of the state senate's version of the bill. There was strong pushback against the proposal in the angling community and its future was unclear.

As for panfish, Wisconsin and Minnesota have acted to limit harvest in an effort to improve the quality of the fisheries. In both states, anglers had expressed dissatisfaction with the size of panfish. The Minnesota DNR cited the widespread use of electronic fish finders and other technologies, along with instant communication about fishing success on social media, as contributors to excessive harvest. In March 2021 the DNR reduced panfish bag limits on ninety-four lakes and streams, and the agency was proposing reductions on fifty more lakes, to take effect in March 2022. Nearly half of the ninety-four waters limit daily bluegill/sunfish harvest to five. Others have a limit of ten; some allow either five bluegills/sunfish and five crappies, or ten bluegills/sunfish and five crappies. Fisheries managers hoped to boost the number of lakes with lower bag limits to about two hundred by 2023, focusing on lakes with biological potential to produce large fish, and where public support for bag limit reductions is strong.

Wisconsin enacted its special panfish regulations in 2016 on nearly one hundred lakes, with similar aims. DNR fisheries staff members will monitor those lakes for ten years to determine which bag limit structures are most effective in producing larger fish. The limits differ among lakes and include:

- 25/10: Keep up to twenty-five panfish per day, but no more than ten of any species.
- 15/5: Keep up to fifteen panfish but no more than five of any species.
- Spawning season 15/5: During May and June, keep up to fifteen, but no more than five of any species; otherwise keep up to a total of twenty-five.

Salting the Water

As the year-round lake populations grow and pressure on lakes increases, a logical question is whether demand for more city-like services will expand. For example, will people used to winter roads plowed and cleared down to bare pavement expect such treatment for town roads in the north, which most often have a cold-weather surface of packed snow with sand applied to the curves and slopes? Bare pavement would require more application of road salt, an emerging concern for lakes in the more southerly parts of the Upper Midwest states.

The issue is serious enough that a group known as the Wisconsin Salt Wise Partnership held the first Wisconsin Salt Awareness Week in January 2021. It offered presentations on the environmental effects of salt, and on ways to use salt responsibly to keep roads and highways safe while protecting lakes, streams, and drinking water. Salt Wise brings together organizations from across the state; it aims to reduce salt pollution by educating residents, government leaders, and winter maintenance professionals and by promoting best salt management practices.

When salt is applied, whether on public roads, on residential driveways and walkways, or on commercial parking lots, it doesn't simply go away. Every ounce of salt put on the ground ends up in the soil and water. Chemically, salt is sodium chloride, a molecule with one atom of sodium and one atom of chlorine. When mixed with rainwater and snowmelt, it dissolves into sodium and chloride ions. When applied near bridges and culverts, the salt can wash directly into a lake or stream. Otherwise, it percolates down to the groundwater and, over time, makes its way to surface waters, to which the groundwater is connected. The closer to waterways salt is applied, the more immediate the risk.

The chloride part of salt is a water pollutant. "Elevated amounts of chloride are toxic to fish, aquatic plants, aquatic insects, amphibians, zooplankton and algae," according to Michigan State University Extension. "If road salt concentrations reach sustained elevated levels, many freshwater organisms will be unable to survive in that waterbody." In addition, salted water, being denser than freshwater, can settle to the deepest part of the lake and accumulate. This chemical layering can prevent natural mixing, leading to a near permanent layer of saltwater in the depths. That layer becomes devoid of oxygen and can no longer support water-dwelling insects and fish.[10]

Once salt is in the water, there is no cheap or easy way to get it out. Salt in lakes, especially in the north, is not yet a serious ecological problem; for example, it is not as immediate a threat as excess phosphorus or invasive species. But in some lakes the chloride content is trending upward. Researchers at the Center for Limnology have compiled and analyzed data on salt in lakes from the early 1980s to 2014. They found chloride levels staying at essentially the same low levels in Northwoods lakes removed from major highways. On the other hand, samples from two lakes along U.S. Highway 51 through north central Wisconsin showed chloride levels rising steadily. Sparkling Lake, a small, clear lake just west of the highway, saw its chloride level increase from just over two parts per million to just under ten parts per million over the three decades. In Trout Lake, a much larger and very deep lake, also near the highway, chloride increased from about 1.4 to about 2.0 parts per million.[11]

These are still low levels. For comparison, the lakes around Madison in southern Wisconsin have chloride levels from just under fifty to just over one hundred parts per million, and trending upward—still not toxic but high enough to raise concern. The EPA sets 230 parts per million as the threshold of ecological concern for chloride in surface waters—that amounts to roughly a teaspoon of salt in five gallons of water. People can start to taste salt in water at about 250 parts per million.[12]

Road salt is the biggest contributor to the salting of lakes, but lake residents can avoid adding to the problem by using salt responsibly on their properties. The Minnesota Pollution Control Agency offers several recommendations:

- Shovel and scrape as much snow and ice as possible; that way less salt will do the job, and sand may suffice to provide traction.
- Don't apply salt if it is 15 degrees F or colder; salt will not melt ice at those temperatures.

- Apply salt sparingly; more salt does not mean more melting. Four pounds of salt is enough for one thousand square feet (an average parking space is about 150 square feet). A heaping twelve-ounce coffee mug full is about a pound of salt.
- Sweep up the excess. If you see salt on dry pavement, it is not helping and will be washed away. Save this salt for later or put it in the trash.[13]

Facing the Stresses

An influx of transient visitors and rising demand for northern lakefront properties point to heavier use of the lakes, more development on shorelines, and more people on the lakes for more days each year. It remains to be seen how, in the long run, these trends will affect lakeshore habitats and lake ecosystems. Will property owners adopt best lake management practices so that scenic values, fish and wildlife habitat, and water quality avoid further impairment, or improve? Or will development proceed so that the northern lakes more closely resemble the crowded, degraded waters found near urban areas farther south? And if the former, what will bring about the change for the better? More protective local zoning ordinances? Legislation at the state level? The hard work of demonstrating to property owners and visitors the importance—social, aesthetic, environmental, economic— of good lake stewardship? Or all of these? That is the subject of the next chapter.

CHAPTER 15

Ways Forward
Forging Connections

The Peacemaker taught us about the Seven Generations. He said, "When you sit in council for the welfare of the people, you must not think of yourself or of your family, not even of your generation." He said, "Make your decisions on behalf of the seven generations coming, so that they may enjoy what you have today."

    —Oren Lyons, Faithkeeper of the Turtle Clan, Onondaga Nation

Advocates for lake protection face substantial challenges in helping to preserve lakes' character and quality. The basic steps necessary are widely known and largely quite simple; that is not the same as easy. Progress is glacially slow and beset by obstacles. For one, leaders in the lakes community see scant prospects for help from state-level policies at a time of tight budgets and deeply partisan and polarized government.

Dave Maturen, president of the Michigan Lakes and Streams Association, observes, "Right now our state legislature is pretty averse to regulations. They have a Regulation Reform Committee, and its focus generally is not so much to add regulations as to do away with them. That doesn't mean we stop trying, and it doesn't mean it won't happen. It's just that in the current climate it's tougher to get new rules and regulations in place." In any case, many existing lake-protective regulations are not rigorously enforced.

That leaves the formidable challenge of encouraging people to do the right things, voluntarily, for the lakes. Those dedicated to lake stewardship know from experience that few converts are won by presenting data and preaching about sustainable practices. Instead, a consensus holds that progress will come from forging and sustaining connections: person to resource, neighbor to neighbor, citizen to legislator, group to group, township to county to state, dollars to needs, and more. Progress also depends on a change in the perception of lakes, lake properties, and our relationships with them.

Starting Point

A vision of a way forward for lakes must first examine the point from which we begin. Lake advocates cite invasive species, nutrient pollution, and unwise and excessive shoreland development as the greatest threats (the latter two intimately connected). A fair question is where the northern lakes now stand in relation to all three. Here the EPA National Lakes Assessment provides some insights.[1] The assessment, performed every five years and most recently in 2017, is a statistical survey of the condition of the nation's lakes, ponds, and reservoirs. Conducted in partnership with state natural resource agencies and Native American tribes, it helps describe how well the lakes support healthy ecosystems and recreation, evaluates the extent of major stresses on lake quality, and helps determine whether lakes nationwide are improving.

The 2012 assessment, the most recent for which detailed information was available as of January 2022, indicated that nutrient pollution (such as from runoff and defective septic systems) was common in the lakes: 40 percent had excessive total phosphorus and 35 percent had excessive total nitrogen. The report noted: "Nutrient pollution is the most widespread stressor measured . . . and can contribute to blooms and affect recreational opportunities in lakes."[2] In addition, the assessment placed one-fourth of lakes "in the most disturbed condition for vegetation along the lakeshore and at the land-water interface. . . . Healthy lakeshore habitat slows pollution runoff and provides varied and complex ecological niches for aquatic life."[3] Wisconsin, Minnesota, and Michigan state agencies provided snapshots of lake conditions based on data they gathered during the 2017 assessment.

Wisconsin Assessment

The overall results of the Wisconsin Department of Natural Resources' 2017 assessment indicated that the state's varied and diverse lakes were mostly in good health, but that nutrient pollution was "a common stressor." The

survey, which looked at about fifty lakes, found 56 percent of them eutrophic or hypereutrophic (high or very high in nutrients), 34 percent mesotrophic (moderate in nutrient content), and 10 percent oligotrophic (low in
nutrients). In terms of phosphorus, the survey found nearly half of lakes to
be in the highest stress categories of most or moderately disturbed.[4]

Michigan Assessment

Michigan's latest report, also covering about fifty lakes, found worrisome
changes in quality between the 2012 and 2017 assessments. It concluded
that "poor lakeshore habitat is the biggest issue" in the state's inland lakes.
In 2012, nearly half of lakes fell into the "most disturbed" category for lack
of diversity in lakeshore habitat and loss of vegetative cover along the shorelines. According to an analysis of the report by Erick Elgin of Michigan State
University Extension, "Lakeshore habitat is considered 'degraded' when too
much vegetation such as standing and downed trees and native aquatic
plants have been removed and replaced by lakefront development, lawns,
man-made beaches and seawalls." The study rated shallow-water aquatic
plants and overhanging brush as "most disturbed" in almost 40 percent
of the lakes. Elgin noted: "Between 2007 and 2012, for every shoreline
indicator, the number of lakes in the 'least disturbed' category (i.e. pristine)
decreased while the number of lakes in the 'most disturbed' category increased. These results may be an indicator that Michigan's inland lakes
need more protections and stewardship."[5]

Minnesota Assessment

Minnesota's survey examined 155 lakes split into three diverse regions: Northern Forests (northeast area, sparsely populated, rich in lakes and woods),
Great Plains (southwest area, largely agricultural), and Eastern Temperate
Forests (a transition zone between the other two). To no surprise, by quality measures such as nutrient level, phosphorus content, algae bloom fre-
quency, and water clarity, lakes in the northeast generally ranked the best
and those in the southwest ranked the poorest. For example, in the Northern Forests region, 80 percent of lakes were rated oligotrophic or mesotrophic; in the Great Plains region more than 95 percent rated eutrophic or
hypereutrophic. Statewide, median water clarity was 7.5 feet in the Northern Forests, 5.2 feet in the Eastern Temperate Forests, and 2.9 feet in the
Great Plains.[6]

Changing Perspectives

Of course, each lake is different; no statistical measure captures the virtues and deficiencies of specific bodies of water. Some lakes are seriously degraded and need remediation where feasible; some are in excellent condition and need only to be preserved in that state; a great many fall somewhere in between. All deserve to be respected and guarded against abuse. But if for the time being that depends on voluntary actions—not changes in government policy—then certainly ways of living on and interacting with lakes need rethinking.

Native American culture brings a different and likely salutary conception of lakes and of water in general. In 2003, Josephine Mandamin (Biidaasige-ba, The One Who Comes With the Light) started the Water Walk movement. An Anishinaabe First Nations grandmother and elder, she walked the perimeter of Lake Superior to raise awareness about abuse and disrespect for water in her home territory in Ontario. She sought to encourage humanity to rekindle a sacred relationship with and connection to water; in the first and subsequent walks, she covered more than 10,500 miles around the Great Lakes, inspiring many others to organize and take part in community water walks. "We've known for a long time that water is alive," she said. "Water can hear you. Water can sense what you are saying and what you are feeling. . . . Give it respect and it can come alive. Like anything. Like a person who is sick . . . if you give them love, take care of them, they'll come alive. They'll feel better. It's the same with our mother, the earth, and the water. Give it love."[7]

Her inspiration took hold in Tinker Schuman (Migizikwe nindizhinkaaz, Eagle Woman), an elder of the Lac du Flambeau Band of Lake Superior Chippewa in north central Wisconsin, who "felt like the Creator was nudging her" to start a walk in her own community.[8] Between 2015 and 2020, she organized a series of annual Water Ways Walks around two lakes on her tribal reservation, which encompasses 260 lakes, sixty-five miles of streams, and twenty-four thousand acres of wetlands.[9] "The purpose is to energize, honor and pray for Mother Earth's lifeblood, water, which brings healing available to everybody," she said.[10] "We're not going to live without this water. Turtles, fish, loons, muskrats, beavers, they all live in the water, and with pollution going on it's not healthy for them or us to live." She sought to raise awareness, not just on the reservation but in surrounding areas, that "it's time to take care of the water."

The walks were held on the Saturday before Mother's Day, starting with a sunrise ceremony at the Bear River Powwow Grounds and a breakfast potluck feast. The walks covered ten miles around Flambeau and Long Interlaken Lakes in Lac du Flambeau. At the beginning, a young girl filled a bucket with water from the Bear River. The water was then passed to a teenage girl, then to a mother, then to Schuman, and after that continuously from woman to woman, "because in Ojibwe culture, the women are life-givers and guardians of the water." From the beginning people of all races took the walks, as did people of all ages, from babies pushed in strollers to an elder pushed in a wheelchair. Says Schuman, "We had surrounding people who would come here from different areas and do the Water Walk with us. *Giganawendaamin Nibi Megwaa Bimoseyang*—'We are protecting water as we walk.' It was really beautiful. . . . Water is essential for life. It is a gift. It deserves our respect, gratitude and reverence."[11]

In another vein, Paul Radomski, a research scientist with the Minnesota Department of Natural Resources, talks in terms of changing long-held conceptions of how lake property owners treat their shorelines. He stresses playing the long game.

We're trying to change behavior, and we know that is very difficult. Look at how long it took us to change our behavior related to drinking and driving. It took fifty or sixty years to change that social norm. It took a lot of effort by many people, and the government wasn't the first to react. It was concerned citizens who said, "This is irresponsible."

In a similar sense, a lawn extending down to the lake has been the social norm. If we can get enough people to become more aware and change their behaviors, then others might just change as well, because they want to fit in and be like everybody else on the lake. That has been the hope: that if we can reach a tipping point, we'll manage to convert most people to a more sustainable management style that's better for lake water quality and for the quality of life. People come from the cities to enjoy the lake, and that means seeing the wildlife, seeing the birds along shore, seeing the loons, watching the warblers come through in spring— all the things people appreciate when they get out in nature.

Protecting Assets

In encouraging property owners to develop their lakefronts more responsibly, Radomski suggests addressing them in terms of their own interests, applying three basic principles.[12] First is to protect their assets—both the quality of the lake and the value of their property. Says Radomski, coauthor of *Lakeshore Living: Designing Places and Communities in the Footprints of Environmental Writers*:

Protect the lake so runoff isn't influencing the quality of the water. Manage the rainfall so it gets into the ground close to where it falls. If you're building a home, think long and hard about where to place it, not only from a rule and regulatory perspective but so that you minimize the impact on the site. A buffer strip of vegetation along the shoreline, as viewed from the cabin, provides a beautiful picture frame for the lake. A naturalized shore also protects the privacy of the cabin. With judicious pruning of trees and shrubs, you can keep the property screened from the view of people on the lake, while still providing an appealing vista of the water.

Second, what value can you bring? It might be adding to the native vegetation so that you do a better job of protecting the shoreline. It could be minimizing the impact of your dock, extending it not too far out so it doesn't create a hazard, and limiting your impact on the near-shore area that is the most sensitive.

The third principle is what Radomski calls connectivity: concentrating activity in a confined area so that the entire lake frontage is not disrupted. That includes creating a single pathway from the home or cabin to the shore, consolidating the pier and boats in one area away from aquatic plants, and limiting the size of the swimming beach.

One-on-one conversations with property owners can be effective. Explains Radomski,

I observe that nobody wishes to degrade their lake. I talk to a lot of lakeshore owners, and they are thoughtful, caring people. As we walk around their lake homes, I see that their desires are often inconsistent with their actions. They mean well, but they haven't been exposed to principles of lake ecology and landscape design. Often, I'll start with a conversation about what they could do to improve their property, for their benefit and the lake's. I'll provide a contact for them to call and discuss the first steps. At the end of the conversation, we talk about the lake's watershed and the land uses within that watershed. In many cases, to protect the lake, they will need to work with their neighbors, their lake association, their

Shoreline restorations like this one can be important in keeping lakes healthy. (Dan Butkus)

community, and multiple government agencies to address pollution reaching the lake from across the landscape.

Water Quality and Property Value

For most lake residents, a direct self-interest lies in the value of their property. Study after study ties the quality of lake environments to property value, and to the vibrancy of community economies and the fiscal health of local governments. That is especially true in lake-rich tourist areas like northern Wisconsin, Minnesota, and Michigan. To cite just a sampling of evidence:

- A Bemidji State University study of 1,205 property sales on thirty-seven lakes in six regions of northern Minnesota found that water clarity was the most important factor in determining lakeshore property values. Professor Patrick Welle, one of the investigators, noted that on average, properties on all lakes in those six regions would rise in value by fifty dollars per foot of lake frontage if water clarity increased by one meter (a little more than three feet). "Now Realtors can talk of location, location, location and clarity, clarity, clarity," noted Welle.[13]
- A University of Wisconsin–Whitewater study, looking at 3,186 real estate sales over ten years on seven lakes in northwest Wisconsin, found that properties on lakes with good water quality had values two to three times higher than on lakes with poor-quality water.[14]
- A Michigan State University study examined the body of evidence on the connection between water quality and property prices. Reviewing forty-three studies, the researchers found that all but two showed a statistically significant relationship: "As a whole, they provide convincing evidence that clean water has a positive effect on property value."[15]
- A study by the University of Wisconsin–Madison found that properties on lakes infested with Eurasian water milfoil experienced an average 7.7 percent decrease in total value.[16]
- A University of Wisconsin–Eau Claire study estimated gains in property value related to water clarity on sixty lakes in the state's Vilas and Oneida counties. It concluded that a one-meter increase in clarity would boost the market price of an average lake home by $8,090 to $32,171.[17]

While these studies do not account for all of the many factors that affect property values, they reinforce the common perception that people choosing lake property care greatly about water quality. Another study in Wisconsin's Oneida County projected the impact on the economy and local government revenue if lake quality were to decline significantly. In that event, according to surveys cited, seasonal residents would spend less time at their properties, and up to 50 percent of visitors would find other vacation destinations. This would cost the county's economy more than $100 million per year; the county would risk losing $2 billion in property value and $19 million per year in property tax revenue.[18]

Voluntary Actions

Many lake residents fail to perceive these connections; even those who see them need advice on how they can do their share. Encouraging property owners to act is a cherished aim of lake associations, state natural resource agencies, and other organizations across the Upper Midwest. Michigan has seen a substantial rise in interest in natural shorelines in the past decade, according to Julia Kirkwood of the state's Department of Environment, Great Lakes, and Energy (EGLE). Kirkwood chairs the Michigan Natural Shoreline Partnership, a collaborative group that promotes ecologically sound lakefront practices. It includes state agencies, Michigan State University Extension, nonprofit groups, and private industry.

Recently, Kirkwood and colleagues have tried to make sustainable shoreland concepts as accessible as possible to landowners. They use the term "lakeshed" to refer to all the land from which water drains into a lake. Says Kirkwood,

> That brings it down to the level of a property owner who is not technically oriented. That's what our whole program is about—providing education and tools in a way that people can understand, so they can assess their own properties and be provided with ideas on how to make some changes. What you do on your land affects your lake. In the next few years, the partnership will be working to get better information, tools, and resources out to help people make that connection and start having good conversations at their lake association meetings rather than focusing all the attention on invasive species or aquatic plant management. What can we do to prevent the eutrophication of our lakes? What can we do to ensure that we still have frogs and salamanders, insects and turtles, and the birds that are supposed to be around a lake? I see more positive energy. More people are getting more information, but it takes a long time. You can't just flick a switch overnight. It will take years to really have a social and cultural change.

One area of emphasis for the partnership centers on alternatives to rock riprap and seawalls for preventing shoreline erosion. That includes creating transition zones between water and land habitat for creatures such as turtles that need to come up onto the shore. Since 2010, the partnership has been educating homeowners and training contractors in how to design and install erosion preventions that use native plantings and natural, biodegradable materials, a concept called bioengineering.

Another key component is Rate Your Shoreland, an online program in which property owners answer questions about their practices in four zones: the near-shore water, the shoreline, the buffer (the strip of land nearest the water), and the upland (the rest of the site).[19] Property owners going through the program can read about why certain practices are harmful or beneficial. At the end, properties can qualify for a Gold, Silver, or Bronze designation. Kirkwood explains,

> We are trying to push the envelope and change the culture of what is truly healthy for the lake. A property that qualifies at the Bronze level is significantly different from a property at the Gold level. Some properties with seawalls can qualify as Bronze; these typically would be larger properties with significant natural vegetation and small structures. We allowed this option because it can be very challenging to remove and replace seawalls, and there are many opportunities on those sites to capture runoff and re-create those near-shore habitat areas. Some properties will never be able to qualify because they are so overbuilt. That's the reality.

Michigan's Cooperative Lakes Monitoring Program also offers a Score the Shore program in which volunteers are trained to assess the status of shoreline habitat on entire lakes. not just individual properties.[20]

In Minnesota, the DNR offers a Score Your Shore toolkit that helps property owners assess the quality of habitat and stewardship on their land and in the adjoining waters.[21] It's designed to show owners what makes a high-quality shoreline buffer. The program includes an illustrated manual that describes the importance of natural shorelines and gives step-by-step instructions for doing a property assessment. Also available are a four-page Quick Guide and a slide presentation for training lake association members as Score Your Shore volunteers.

Wisconsin has a Shoreland Evaluation Tool that walks users through questions about the management of the shoreland zones of their property.[22] The

responses dictate whether the shoreland is at the "entry, mitigation, restoration, or protection" level. Users can review their answers, receive ideas
for improvement, and upload pictures of their property. They can also save
their results and come back after deploying best practices to see how their
rating has improved. The tool directs property owners to information about
best practices under the state's Healthy Lakes and Rivers program. Lake
organizations can set up their own pages within the tool.

Lake Association Initiatives

Some lake associations have brought shoreland evaluation concepts at various levels directly to their members. Their experiences provide glimpses
of what promotion of lake stewardship can accomplish, of the challenges
that can impede progress, and of innovative ways to enlist property owners'
participation.

Tomahawk Lake, 3,300 acres of state-designated Outstanding Resource
Water in northern Wisconsin's Oneida County, has an estimated 49 percent of its 30.2-mile shoreline in public ownership; several large, privately
owned parcels also remain in a mostly natural state. "That is one reason
we have a high-quality lake," says James Kavemeier, chair of the Tomahawk
Lake Association's environmental and education committee. "We want to
keep what we've got, and we want to make sure property owners know that
their shoreland practices can have an impact on water quality." To that end,
the association embarked on a project to identify and protect areas of critical habitat along the lakeshore, while also encouraging property owners to
follow best practices on their properties.

The critical habitat project was conducted by Michael Meyer, owner of
NOVA Ecological Services and a retired Wisconsin DNR research scientist,
along with colleague Jim Kreitlow, a retired DNR lakes biologist. Working
under a DNR lake planning grant acquired by the association, Kreitlow and
Meyer identified and mapped thirty-four critical shoreline areas including
fish and wildlife habitat, significant wetlands, and important aquatic plant
communities. For each one, they created a narrative with photographs, an
explanation of its importance, and recommendations for its protection. The
work also included boat-based surveys to identify features such as pieces
of submerged wood (more than four thousand located) and areas of rock
and rubble bottom suitable for walleye spawning. In addition, all 414 properties fronting on the lake were surveyed for evidence of erosion and runoff, the presence and condition of any vegetated shoreline buffer, and an

accounting of piers, boats, and other items stored near the shoreline. The surveys followed established Wisconsin DNR protocols. "We put it all together on a GIS map and overlaid that on the critical habitat areas," says Meyer. "The idea is to help the association target education and outreach to landowners who have shoreland issues and are in close proximity to the critical habitats."

The work was completed in March 2020. Nearly concurrent with that work, Kavemeier came upon the Michigan Rate Your Shoreland program and through various forms of outreach encouraged association members to complete it. Fifteen property owners did so; each then received an offer of a site evaluation. That led to a handful of completed projects: a rock infiltration bed to capture runoff, a couple of natural shoreline plantings, a few erosion-control projects. "It's small steps," says Kavemeier.

He believes the lake's generally good condition may actually be a deterrent to action. Control of Eurasian water milfoil, rather than water quality, tends to dominate the attentions of residents and the lake association. Another deterrent is that once property owners understand the importance of natural shorelands, they run into difficulty getting work done. Kavemeier observes, "Say I'm an ordinary homeowner. I go to my local nursery and say, 'I want to restore my shoreline and I would like you to recommend plants.' There is limited expertise among local nursery people." In addition, he says, many people are intimidated by the process of applying for cost-sharing grants to improve their properties; others simply may prefer manicured lawns.

Kavemeier sees a need to reach out not just to people interested enough to request site visits and attend natural shoreland workshops, but also to all those who simply enjoy the lake: "We know through our survey that people are interested in water quality and wildlife habitat, but they don't have an understanding of what to do, and they may not know what the best shoreland management practices are." Accordingly, Kavemeier's committee in early 2021 opted for a simpler approach: recognizing people who take or pledge to take five basic steps:

- Capture water runoff before it reaches the lake.
- Establish a no-mow area along the shoreline or create a buffer of native plants and trees.
- Minimize use of fertilizers, pesticides, and herbicides.
- Maintain the septic system.
- Allow native aquatic plants and fallen trees to remain in the lake.

Those who agree will be recognized in the association newsletter as Lake Stewards. The association updated its website to provide information about the five steps, how to complete them, and why they are important. Kave-meier says. "I'm beginning to think that something is better than nothing: If we celebrate small gains, we might make big gains in the long run."

Seeking Simplicity

A simplified approach is also taking hold in Minnesota. After retiring from her medical practice, Dorothy Whitmer became active in the Gull Chain of Lakes Association (GCOLA) in Minnesota's Crow Wing and Cass counties. Now a Lake Steward program she helped launch for the chain of eight lakes (thirteen thousand total acres) is being rolled out to lake associations across the state by Minnesota Lakes and Rivers Advocates. It recognizes property owners who maintain buffers of trees and plants along their shorelines; keep at least half the rest of their land in native trees, shrubs, and vegetation; refrain from using broadcast fertilizers and pesticides; and have no rock riprap on their shoreline—or if they do, allow plants to grow in it.

Whitmer was inspired to act after she took a Michigan's Rate Your Shore-land survey: "I took the quiz thinking, 'I'm going to get the best score in the world. I'm a great lakeshore owner and I've got this wonderful property.'" But her score earned her only a Bronze rating. "I was very upset about that, because I was doing everything wrong. I had a lawn; I was using fertilizer and pesticides on it. I had let part of my buffer zone grow, and I got all these native plants, but it was only part of my lakeshore. I felt my performance was abysmal."

Around that time, at a GCOLA meeting, she asked a state DNR represen-tative what she and her neighbors should do to improve their shorelines: "He said, 'Quit mowing.' So my husband and I put a stake out twenty-five or thirty feet from the water and said, 'We're not going to mow here.'" Then she wondered how to encourage other property owners to take simi-lar measures. She had seen attempts at outreach and education fail: "The knee-jerk response is, 'Let's go educate these people.' My county has a glossy booklet for shoreland owners telling them what they've got to do. You hand somebody a booklet, and right away it's in the trash. You have to meet people halfway at least. You need lakeshore owners to do something to come for-ward. The something we came up with was a quiz."

The chain of lakes has about two thousand properties, and about 960 owners are GCOLA members. Whitmer worked with Kris Driessen and

Sheila Johnston, who lead the association's Shoreline Restoration Program, to develop a simple four-question quiz. They sent an email to the membership announcing a Lake Steward program aimed at improving water quality, and inviting members to take the quiz. A link directed recipients to a Score Your Shore Mini-Assessment. "We sent it out on Friday of Memorial Day weekend in 2019," says Whitmer. "It rained on that weekend, so people were in their lake homes on their laptops." Within a few days, about a hundred people responded. To those, Whitmer emailed an offer of a visit to evaluate and score the property for Lake Steward recognition. "We ended up doing visits to twenty-eight properties scattered all around Gull Lake," she says. "Some scored very high, and some scored abysmally, but it didn't matter. We wanted to see them all. Out of those twenty-eight site visits, we awarded twenty-three Lake Stewards."

Each Lake Steward receives a colorful twelve-by-eighteen-inch aluminum sign. "People are proud to put it on their docks," says Whitmer. "They passionately want that award." She looks at Lake Steward as a kind of foot in

Three generations of Lake Stewards on Minnesota's Gull Lake (from left), Theresa Whitman, Taous Khazem, and her son Layth Yabdri. (Sheila Farrell Johnston)

the door to start people thinking more deeply about lake protection. Parts of four cities front on Gull Lake, which is served by sewers. Manicured lawns are common. Whitmer observes,

> The mindset is that we need those lawns, because that makes us look good. We associate that with being admirable and being good people. But no one is born wanting a lawn. If you want to know what's innate, it's birds. It's the wind through the trees. That's what we want to get back to. That's why people go to the lakes. We have a situation where humans want the natural world and yearn for it, but they're doing the exact opposite. They're shooting themselves in the foot and ruining their own property values. That's a situation where we can intervene. If we can get them to do the quiz, then we can go out and start talking.

And that may lead to actual shoreland restorations, for which the association provides cost-sharing grants. As of March 2021, GCOLA had awarded fifty-one Lake Stewards encompassing two miles of Gull Lake shoreline, and twenty-six property owners had inquired about shoreland projects.

"The real success is when we convert someone who has a lawn," says Whitmer. "The people who are truly naturalists are the ones who set an example. You put a Lake Steward sign in front of a forested shoreline and people take notice. We have a map of Gull Lake with all the Lake Steward properties, and they cluster, because neighbors talk to neighbors."

Minnesota Lakes and Rivers Advocates learned about the Lake Steward initiative and in December 2020 invited its 256 lake associations to pilot-test it; thirty quickly agreed to do so, according to Jeff Forester, executive director. His statewide association uses a longer but still simple ten-question evaluation. Training for property evaluators was held in April 2021. Says Forester,

> We're taking the thirty-thousand-foot view initially. The survey covers what we find to be the ten most critical issues. Obviously we could go much deeper, but you have to start somewhere. Now that we've got a platform, we're going to keep educating people. What does woody debris really do in a lake? Why is it important? Why are native plants more beneficial than nonnatives?
>
> As we start getting the Lake Steward signs out, we hope this thing will snowball. I see it as a way to get people interested. Then we can start

giving them information they need to educate themselves, their neighbors, and people who come to visit them. People will tell the story: "We had to restore the shoreline because runoff was going down into the lake." So instead of someone with a green clipboard telling you what to do and making you feel like a dummy, you're talking with a neighbor or a friend who's proud of what they did. That's why this idea is so brilliant.

A Toolkit for Lake Protection

There is more than one way to propagate lake stewardship initiatives across multiple lake groups. A Wisconsin advocacy organization is developing a Lake User Toolbox with the aim to help groups identify and mitigate threats to lake resources without having to create lake management plans, which many cannot afford. Wisconsin's Green Fire, a nonprofit that promotes science-based natural resource management, is developing the toolbox with NOVA Ecological Services under a foundation grant.

Bob Martini, a Green Fire board member, explains:

Most people want to protect their lake, but they don't know what to do. Our idea is to create a simplified way for lake associations to evaluate threats in their watershed, so that they can be more specific in their communications with landowners. It's one thing to say, "It's good to have shoreline buffers." Everybody probably has heard that kind of advice before. It's another thing to say, "On this particular stretch of shoreline, we have walleye spawning," or "We have this kind of bird nesting in this area." Then once you know which parcels have problems, or may have problems in the future, what can landowners do about them? They would be under no obligation to do anything, but at least they'd know what the optimal approaches would be if they did decide to take action.

When completed, the toolbox will enable lake and watershed groups to proceed in three phases. First, assess current impacts on the lake or watershed. Second, identify specific threats to water quality and fish and wildlife habitat at the level of individual properties. Third, set priorities for steps property owners could take to ensure that the lake is protected. "This is a process that doesn't have to get into a grant cycle," says Martini. "It doesn't require a consultant, and it doesn't cost a large amount of money." Toolbox users could draw from a variety of existing databases in making the initial lake or watershed assessment, and then proceed to the later phases. The databases cover information such as satellite mapping, soils, vegetative cover, lake chemistry, nutrient status, and the lake water source, and much more.

"What we envision is an intermediate approach between doing nothing and opening a full-blown DNR management grant process, which a lot of groups find daunting," says Martini. "This is something they can do on their own. The main function is to help people understand in more detail what the threats are in the watershed and then convey that to individual landowners." After using the toolbox, a group still could pursue a lake management plan and apply for grants; if so, they would come to the process having shown a commitment to the resource and so with a potential advantage in competing for funds.

Meyer, of NOVA Ecological, sees the program including a series of teaching workshops covering basic lake ecology, how to gather lake-specific information from DNR databases and county and university extension sources, and how to synthesize and use the information collected. "The goal is to find the sweet spot: What information can be obtained at no cost?" says Meyer. "There's a lot of information out there for people who know how to go after it." Even data that is not lake-specific can be useful: "Is it a seepage lake, a drainage lake, or a drained lake? What does that tell you about the watershed? What might you need to worry about given your lake type? It's about helping people think about their lakes in a more comprehensive way."

Care for the Fishes

For lake advocates concerned specifically with fisheries and fish habitat, one resource is the Midwest Glacial Lakes Partnership. This group of more than a thousand resource professionals has a mission to protect, rehabilitate, and enhance sustainable fish habitats for today and for future generations. Among its offerings is an annual grant program for fisheries habitat conservation projects. "In broad strokes, the grants are designed to fund conservation work that first and foremost benefits inland lake fishes," says Joe Nohner, partnership coordinator. "That can be in the watershed, in the lake, or on the shoreline."

Grants have funded a wide range of conservation projects to benefit imperiled and endangered fish species as well as those favored for sport and recreational angling. Typically, three to five projects are funded each year for $10,000 to $75,000; larger projects up to about $300,000 can be considered. Projects can encompass one or more sites, lakes, or watersheds and address the causes of fish habitat impairment as opposed to managing symptoms. Some examples of project areas include water quality and erosion control, lakeshore or in-lake habitat restoration, enhancement of fish migration passages, evaluation of habitat conditions, climate change adaptation or mitigation, and training of biologists on inland lake fish habitat management tools and approaches. "We also invest in education, outreach, and communication," Nohner says. "Ultimately, conservation comes back to working with people on lakes with substantial private ownership. You're convincing people to do what's right for the lake."

Permanent Preservation

While lake associations look to improve shorelands one small parcel at a time, land trusts and land conservancies aim to shield larger tracts of high-quality lakeshore habitats from development. Many of these nonprofit organizations operate in the Upper Midwest and across the United States. They typically deal only with parcels of a certain minimum size and with special ecological, scenic, or conservation value. They permanently preserve lands and shorelines that otherwise could be sold privately and subdivided for homes, cottages, condominiums, or other purposes. Lands can be protected in various ways:

Gifts of Land

Conservancies and land trusts accept land donations that, depending on the donor's wishes, can be managed as nature preserves or working forest reserves. Sometimes these properties are opened to the public.

Purchases

In special cases, a land trust can buy properties outright. Sometimes the owners sell the property at less than fair market value, making it more affordable for the nonprofit organization to acquire, while giving the owner tax benefits.

Conservation Easements

Here the owner does not donate or sell the property but enters a legally binding agreement to protect its conservation values. In turn, the land trust agrees to uphold the terms of the easement in perpetuity. Easements can be tailored to the owner's wishes. For example, owners can continue to live on the property and undertake activities like harvesting timber and cutting trails, so long as the essential values of the land are preserved.

Deed Restrictions

These provisions legally require future owners of the property to preserve it in a natural state. While deed restrictions historically have been difficult to enforce over the long term, a land trust or conservancy can fulfill that role; it has standing to see that the conservation restrictions are observed.

A good example of a land trust's success was the 2018 donation of 430 acres of woodlands, wetlands, and wildlife habitat and 4.4 miles of natural shoreline on Lake Katherine in Wisconsin's Oneida County to the North-woods Land Trust. It was one of the largest outright donations to a land trust in the state and possibly the largest private donation of lake front-age to a land trust in Wisconsin history. Now called the Yawkey Forest Reserve, the land was donated by Yawkey Lumber Company; it had been under the same ownership for 125 years, according to company president Robert Hagge Jr. Since 2003, the Northwoods Land Trust has permanently conserved 14,053 acres of land with ninety-two conservation easements and twenty-five purchased properties. These acquisitions have protected forty miles of lake shoreline and thirty-five miles of river frontage from development.[23]

The Policy Front

Ultimately, lake protection depends in part on government policy; some lifts are too heavy for lake groups and property owners to undertake on their own. Leaders of the state lake associations in Michigan, Minnesota, and Wisconsin remain active on the policy front despite a discouraging political climate, and all three promote variations on forging connections as a key to progress.

Reaching across State Borders

Dave Maturen, president of the Michigan Lakes and Streams Associa-tion, sees a need for state help with aquatic invasive species control and prevention.

> There are not enough resources at the state level. All the surface water in Michigan is theoretically in control of the state under the Public Trust Doctrine. We need the state to step up and recognize its responsibility in controlling AIS, instead of leaving it up to the residents of the lakes. We need more active patrolling. We need more boat wash stations. Many

lakes take care of AIS with special assessment districts, so maybe state officials think they can wash their hands of it. It's great to have the locals realize that AIS affects their quality of life and their property value, but in my opinion the state has a big stake in it, and they ought to step up from a monetary or regulatory standpoint.

He's also concerned about emerging contaminants, notably per- and poly-fluoroalkyl substances (PFAS) that have made their way into rivers and the groundwater on which many lake residents rely. PFAS, used in flame retardants and a wide range of household products like nonstick pans, carpets, and waterproof clothing, are extremely persistent chemicals that have been linked to cancer and various other health problems.

Maturen takes pride in his organization's partnerships, developed over decades, with the Michigan DNR and EGLE, the Michigan Inland Lakes Partnership, the Michigan Environmental Council, the Michigan Natural Shoreline Partnership, Michigan United Conservation clubs, and others: "All these groups are focusing on the issues that are really germane to our mission. We also talk to groups from Maine to Oregon that are working on the same issues. The relationships aren't as formal, but they are certainly valuable. There's not a heck of a lot of difference in land use and lake issues no matter where you go in this nation. We can learn from them and share data, studies, and information. We can explore what strategies they are using, what has worked for them and what has not."

The group is also an essential partner in the state's Cooperative Lakes Monitoring Program, in which volunteers regularly check their lakes for water clarity, temperature, phosphorus, and other parameters (Wisconsin and Minnesota have similar programs run through their state natural resource agencies). "All data collected goes to state agencies so that residents can look at what has happened to their lakes over the decades," Maturen says. "Is something happening in your lake that's not happening in the other lakes in your county? Is it something upstream? We have a lot of committed people who come back year after year and do the monitoring."

A Civic Action Blueprint

Minnesota Lakes and Rivers Advocates takes a different approach to connections, advocating a framework known as civic governance that pulls together a broad range of interests. Jeff Forester, executive director, explains:

Typically a lake association will work with the local watershed district. They probably know the DNR fisheries specialist, and maybe they know the county resource person. But they're not interacting with the Chamber of Commerce. They're not associated with any faith communities. They're not plugged in with the Girl Scouts, the Boy Scouts, the 4H, or the FFA. They're not connected to the farmers, the agricultural co-op, the real estate community. And yet all of those have a stake in keeping the lake healthy, because lakes are really big economic engines in their areas. We're saying, let's bring everybody together by organizing around shared values.

So the lake association wants to protect the lake to preserve a way of life. The real estate agent across the table wants to protect the lake so as to drive higher property values and earn more money. The city needs the tax base. Maybe the school district would like to set up a classroom on the lake. It's a Venn diagram of where the different interests overlap. That's the point at which active citizenship takes place. If we're all working on water quality for the community, for the public good, we need to know everyone's personal agenda, so we can craft solutions that benefit everyone.

Forester contrasts that approach with the typical scenario of "white hot" conflict around environmental issues. Two opposing sides mobilize and engage in a struggle. "Both sides are yelling at each other," says Forester. "At the end of the exercise both groups are damaged, and they have spent all their political capital. Screaming at each other burns power. It does not create it." On the other hand, when issues are resolved cooperatively, relationships are built and all involved end up with more power and political capital: "If we want to protect lakes, then the people who love lakes need more power. We're trying to reimagine what citizenship means and how it is expressed, starting with local communities and based around water resources."

Forester cites the Minnesota Clean Water, Land, and Legacy Amendment of 2008 as an example of what can happen when diverse interests unify around a public good. An amendment to the state constitution established a 0.385 percent sales tax with designated shares for enhancement of wetland, prairie, forests, and habitat for fish, game, and wildlife; protection of lakes, streams, and groundwater; funding for the arts and cultural heritage; and support for parks and trails. The process started, says Forester, because pheasant hunters wanted more habitat for their favorite quarry and duck hunters wanted to protect the waterfowl sloughs in western Minnesota. Those constituencies were too small to move policy at the state level, but when the water, parks, and arts interests were brought into the picture, the measure won approval. From 2010 through 2020, the tax raised nearly $3.3 billion for the four purposes.

Uniting around Water

Michael Engleson, executive director of Wisconsin Lakes, advocates for building a foundation to prepare for the prospect of a more hospitable political climate in the future. He already has seen some limited policy consensus around sustainable farming practices that could benefit surface and groundwater in some areas of the state. He supports a continued focus on key initiatives like sustainable lakeshore development and invasive species prevention "so that they don't get lost when the time comes where we can get some positive change to happen." At the same time, he suggests setting an expansive vision for the future of lakes, even if it's a "pie in the sky" agenda that might never become reality: "We need to look at where we want to get in the long run, because even getting partway is going to be a big help."

He envisions forming close ties with water-related interest groups, including those without obvious connections to lake health. That means groups concerned with flooding issues statewide, with lead in household drinking water services, with urban environmental justice issues, with groundwater contaminants like nitrate and PFAS, with climate change mitigation. "Sometimes I feel like the lakes community has our own silo that we go down," Engleson says. "We end up actually competing against these other interests, when if we all worked together and thought of water as a general issue, we would end up having more success. We'd certainly put ourselves in a position to be more successful when the politics loosen up and some things start happening."

Engleson sees the state's hundreds of lake associations and lake districts as essential allies with strong and dedicated memberships and volunteer capacity to protect and improve their waters: "They have done a really good job of promoting sustainable shoreland practices, getting property owners to understand the importance of shoreline buffers and natural vegetation. I've seen lake districts create funds to provide small grants to help homeowners install shoreland protections. There are lots of examples across the state of lake groups banding together at the county level to do good things."

Some districts have worked effectively with farm groups in the watersheds to help reduce phosphorus inputs to their lakes, he says.

"If there has been a limitation so far, it's that they are often very inward looking," Engleson observes. "They want to do the work on the water that's right in front of them. They could benefit from being more communicative with non-lake organizations in their own areas. That could include forging partnerships with fishing and hunting groups, and even with service groups in communities associated with the lake, thus bringing non-waterfront property owners into the conversations. That is the next step where lake groups need to go."

Bob Martini, of Wisconsin's Green Fire, also argues for more diverse collaborations.

> Groups involved in fishing, boating, and silent sports are often more motivated to protect the lakes than landowner groups are, and yet they've been underrepresented in the lakes discussion. If you look at the past thirty years of lake-related conventions, you see very few user groups and a whole lot of landowner groups represented. The users feel kind of powerless. They go to many lakes in a given year instead of just one. They see the lakes changing. They see the problems firsthand. They are an untapped resource.
>
> It's analogous to the food system. Most of the agriculture programs deal with landowners, the farmers, but people who eat the food are now starting to exercise some influence. They're looking for fewer additives, less genetic engineering, more land protection, more nutrition, and more organics. We don't have that with lakes. We need to evolve and encourage lake users to apply pressure for change, whether voluntary or through regulations. User groups that are large enough in the voting population can make some inroads.

Meanwhile, Engleson and Wisconsin Lakes colleagues are looking to help the lakes community and especially lake associations to build political muscle through a Lakes Action program. It includes workshops and webinars that teach the techniques of advocacy and influencing legislators: "We want the lakes community to be known within the capitol as a force to be listened to, just like the agricultural and business communities and other interest groups. Having the lakes community become a known entity could be really powerful."

Setting Priorities

Taken together, the pressures on lakes foretell gradually declining quality; to protect them, governments and lake advocacy groups need to decide where to invest resources. Martini observes, "Some environmental impacts are reversible, like the pollution of a river. Some are not, like phosphorus pollution of a lake or climate change. We need to work harder on the latter, but historically we have concentrated on the former, probably because it's easier."

An analysis by Paul Radomski and Minnesota DNR research colleague Kristin Carlson offers further insights. "We predicted a six times greater return on investment by protecting high-quality lakes than focusing on impaired lakes," says their article's abstract. "We suggest that policymakers re-evaluate the distribution of those funds and that they consider investing a greater percentage to protect lakes at risk before they become impaired."[24]

The article notes that traditionally, about four-fifths of Minnesota's lake quality restoration and protection funds have gone to rehabilitate impaired waters, and one-fifth to protect unimpaired, high-quality lakes that are at risk. The study looked at data on more than 2,700 lakes; it included a cost-benefit analysis of actions that could be taken to improve and protect the waters. The researchers noted that restoring degraded lakes could yield a good return on investment for lakes with high-value land around them, and on lakes in or near cities that are used heavily for recreation. At the same time, they found drawbacks in focusing too much on impaired waters. In some lakes, phosphorus loading—a major cause of impairment—comes from nonpoint sources like farm runoff that are challenging and costly to control. In addition, excess phosphorus already in the lake can limit prospects for successful restoration.

Radomski observes,

Lakes have memory, and it's called sediment. Once the sediment is loaded up with nutrients, it has the potential to cycle. In the cities, we do alum treatments to bind the phosphorus to the sediment, but that is short-term and very expensive. Are we getting the best bang for our conservation dollar? Why aren't we spending more money on lakes in northern Minnesota, where we still have some very close to pristine water-quality conditions, but where there will be very high development stress in the near future? We could allocate our dollars to put more into lake protection. It's a big-dollar question: Where is it appropriate to spend? If you put all the resources toward impaired waters, it's quite a waste of money. Once a lake reaches a tipping point, it is really hard to fix.

Residents of impaired lakes may have different ideas on this topic; balancing priorities for lake repair and lake conservation will surely challenge resource managers.

Doing What Is Essential

To anyone deeply involved in lake advocacy, the challenges facing lakes can appear daunting. When asked, state lake association leaders suggested focusing on these essentials:

- Keep or create sustainable shorelands. "Do the buffers, build rain gardens, and limit hard surfaces so that the water slows down before it gets to the lake," says Engleson.
- Take advantage of grants. State natural resource agencies offer a variety of grants to help pay for shoreland improvements, invasive species prevention, and other projects.
- Seek expert advice. Maturen observes, "There are a lot of consultants who do lake studies, cleanups, monitoring, and remediation. Lake groups should know they can hire one of these entities to look at their lake and provide a new perspective."
- Join the lake association. "The old adage says that in union there is strength," says Maturen. Engleson adds: "Volunteer. Learn the issues on your lake more broadly, not just on your own shoreline."
- Communicate with policymakers. "We need policies to fix the big problems," says Engleson. "The only way that's going to happen is if legislators understand that this is important."

Epilogue

Toward a Land and Water Ethic

I like to say I live on 180.5 acres. The 0.5 is the wooded lot on which our modest house stands. The rest is Birch Lake; it's the reason my wife and I live where we do. Of course, we don't own the lake, but we treasure it and try to care for and protect it as if we did. Why would we do things on our 0.5 acres of land that would harm the 180 acres of water that nurtures an abundance and diversity of life and, for that matter, our souls?

Perhaps in that question lies the essence of what lake stewardship means. In living on and with our lakes, we need to think beyond ourselves. About the lake as a living, breathing entity. About our neighbors. About the people who visit for all manner of enjoyment, not the least being fish and wildlife, scenic beauty, quiet serenity. About our children and grandchildren, who, we hope, will enjoy the lakes long after we are gone.

Surely in that conception of lake life one can find an echo of the land ethic advocated by the late naturalist Aldo Leopold, who envisioned humanity not as separate from or dominant over the natural world but as an intimate player in it. Because Leopold's specialty was wildlife ecology, many think of his principles in terms of the terrestrial world. However, his land ethic, and the most concise summation of it, apply equally to the waters:

> A thing is right when it tends to preserve the integrity, stability and beauty of the biotic community. It is wrong when it tends otherwise.

Anyone who reads Leopold's signature work, *A Sand County Almanac,* can find fertile soil in which a personal land and water ethic can grow.

Historically, as a society, we have treated our waters abominably, far worse than we have treated the land. Specifically, we filled out rivers, lakes, and oceans with all manner of pollutants and filth. We mostly stopped doing that after the first Earth Day in April 1970, and after the Clean Water Act of 1972 established the goal, largely achieved, of fishable and swimmable waters. Things are much better now. And we also have much work left to do.

Too often we think of our lakes as surfaces on which to enjoy our favored forms of recreation; we forget that they are complex ecosystems ruled by laws of biology, chemistry, and physics working in balance. If in the world of Sir Isaac Newton every action has an equal and opposite reaction, then in the world of lakes any action we take, on the water or the land, can have a repercussion, for good or ill. As lake residents and lake users, we have a collective duty to see that our actions, in the aggregate, are for good.

What that actually means is perhaps deceptively simple: Step lightly. Live gently. Take with discretion. Give back in abundance. Sometimes the most basic watchwords are the most worthy of sharing. Sometimes to one's frustration they are also readily ignored. Often the path of lake preservation seems to lead relentlessly uphill. It is useful at such times to remember the words of the late anthropologist Margaret Mead:

> Never doubt that a small group of thoughtful committed citizens can change the world; indeed, it's the only thing that ever has.

If that's true, then surely a group of thoughtful, committed lake advocates, tens of thousands strong, can work wonders.

Notes

Chapter 2. Cabin Country

1. Wisconsin Department of Natural Resources, Division of Forestry, "Appendix B: Cultural History of Wisconsin's Forests," in *Wisconsin Statewide Forest Assessment 2010*, 400–404, https://dnr.wisconsin.gov/sites/default/files/topic/ForestPlanning/WisconsinAssessment_final_061810.pdf.

2. John T. Curtis, *The Vegetation of Wisconsin: An Ordination of Plant Communities* (Madison: University of Wisconsin Press, 1959), 219–20.

3. Wisconsin Department of Natural Resources, "Appendix B: Cultural History of Wisconsin's Forests"; Robert Gough, *Farming the Cutover: A Social History of Northern Wisconsin, 1900–1940* (Lawrence: University Press of Kansas, 1997).

4. John Bates, *Our Living Ancestors: The History and Ecology of Old-Growth Forests in Wisconsin and Where to Find Them* (Mercer, WI: Manitowish River Press, 2018), 32.

5. Bates, *Our Living Ancestors*, 40.

6. "In 20 More Years, Minnesota's Giant White Pine Forest Would Be Entirely Cut," Forest, Fields, and the Falls, Minnesota Historical Society, accessed January 2020, http://www.mnhs.org/sites/default/files/forestsfieldsfalls/lumbering/lumbering-slide08.html.

7. Bates, *Our Living Ancestors*, 40.

8. "Michigan Forests: The Beginning of Conservation," Michigan State University, accessed January 2020, http://uptreeid.com/History/ConsEra.htm.

9. Joseph J. Jones, "Transforming the Cutover: The Establishment of National Forests in Northern Michigan," *Forest History Today* (Spring/Fall 2011): 49.

10. Aaron Shapiro, *The Lure of the North Woods: Cultivating Tourism in the Upper Midwest* (Minneapolis: University of Minnesota Press, 2013), 20.

Chapter 3. Paradise Discovered

1. The real estate listings interspersed in this chapter (from the *Lakeland Times* newspaper in Wisconsin's Oneida County) generally show the rate and magnitude

of rising lake property prices from 1980 into the 2000s. The properties selected may not be strictly comparable; prices vary with the quality of the lake, beach frontage, structure condition, and other factors.

2. Greta Kaul, "The Uncertain Future of Cabins in Minnesota," MinnPost, April 27, 2018, https://www.minnpost.com/economic-vitality-in-greater-minnesota /2018/04/uncertain-future-cabins-minnesota.

3. Greta Kaul, "Last Resorts: The Slow Disappearance of a Northern Minnesota Tourism Institution," MinnPost, March 26, 2018, https://www.minnpost.com/eco nomic-vitality-in-greater-minnesota/2018/03/last-resorts-slow-disappearance -northern-minnesota-tourism-institution/.

4. "Annual Minnesota Resort Sales Tax Statistics," Explore Minnesota, November 4, 2021, https://mn.gov/tourism-industry/research/reports.jsp?nid=1439.

5. Kaomi Lee, "Mom and Pop Resorts in Minnesota Are Disappearing," TPT Originals, July 19, 2018, https://www.tptoriginals.org/mom-and-pop-resorts-in-min nesota-are-disappearing.

6. Kaul, "Last Resorts."

7. Lynn Markham, "Shoreland Development Density and Impervious Surfaces," Center for Land Use Education, University of Wisconsin–Stevens Point, 2003, https://www.uwsp.edu/cnr-ap/clue/Documents/Water/Shoreland_Development _Density_and_Impervious_Surfaces.pdf (statistics drawn from "The Northern Lakes and Shorelands Study," Wisconsin Department of Natural Resources, 1996).

8. "Minnesota Lakes and Rivers Advocates [MLR] Lake Home and Cabin Ownership Study," October 2016, https://mnlakesandrivers.org/wp-content/uploads/ 2021/11/Public2016MLRStudyResponses.docx.pdf.

9. Kaul, "The Uncertain Future of Cabins in Minnesota."

Chapter 4. One Water

1. "Wisconsin Water Facts," Wisconsin Water Library, accessed March 2020, https://waterlibrary.aqua.wisc.edu/water-facts/.

2. "Groundwater Facts," National Ground Water Association, accessed March 2020, https://www.ngwa.org/what-is-groundwater/About-groundwater/groundwa ter-facts.

Chapter 5. The Trouble with Phosphorus

1. Kaye LaFond, "Harmful Algal Blooms Showing Up in Northern Michigan Lakes," Michigan Radio, October 18, 2019, https://www.michiganradio.org/post/ harmful-algal-blooms-showing-northern-michigan-lakes.

2. "Six Mile Lake," Tip of the Mitt Watershed Council, accessed April 2020, https://www.watershedcouncil.org/six-mile-lake.html.

3. Linda Gallagher, "Blue-Green Algae Discovered in Area Lakes," *The Review* (covering Antrim and Kalkaska Counties), September 12, 2019, http://www.antrim review.net/content/blue-green-algae-discovered-area-lakes.

4. Ron Struss, "The '500 lbs. Algae Adage': Where Did It Come From?," *Lake Tides* 28, no. 3 (Summer 2003): 1.

5. Struss, "The '500 lbs. Algae Adage,'" 1–2.

6. LaFond, "Harmful Algal Blooms."

7. Erik Ness, "The Model Lake," *Isthmus*, August 24, 2017.

Chapter 6. Changing Lakescapes

1. See also John C. Panuska and Richard A. Lillie, "Phosphorus Loadings from Wisconsin Watersheds: Recommended Phosphorus Export Coefficients for Agricultural and Forested Watersheds," *Research Management Findings* 38 (April 1995).

2. Lynn Markham, "Shoreland Development Density and Impervious Surfaces" (University of Wisconsin–Stevens Point, Center for Land Use Education, 2003), 11.

3. Michael Meyer, James Woodford, Sandra Gillum, and Terry Daulton. *Shoreland Zoning Regulations Do Not Adequately Protect Wildlife Habitat in Northern Wisconsin*, Final Report—USFWS State Partnership Grant P-1-W, Segment 17 (Bureau of Integrated Science Services, Wisconsin Department of Natural Resources, 1997).

4. Alec R. Lindsay, Sandra S. Gillum, and Michael W. Meyer, "Influence of Lakeshore Development on Breeding Bird Communities in a Mixed Northern Forest," *Biological Conservation* 107 (2002): 1–11.

5. Jeffrey Reed, "Influence of Shoreline Development on Nest Site Selection by Largemouth Bass and Black Crappie" (poster, North American Lake Management Conference, Madison, WI, 2001).

6. James E. Woodford and Michael W. Meyer, "Impact of Lakeshore Development on Green Frog (*Rana clamitans*) Abundance," *Biological Conservation* 110, no. 2 (April 2003): 277–84; Meyer et al., *Shoreland Zoning Regulations*.

7. S. J. Kerr, B. W. Corbett, N. J. Hutchinson, D. Kinsman, J. H. Leach, D. Puddister, L. Stanfield, and N. Ward, *Walleye Habitat: A Synthesis of Current Knowledge with Guidelines for Conservation* (Percid Community Synthesis Walleye Habitat Working Group, Ontario Ministry of Natural Resources, 1997); George C. Becker, *Fishes of Wisconsin* (Madison: University of Wisconsin Press, 1983); Amy L. Leis and Michael G. Fox, "Effect of Mine Tailings on the In Situ Survival of Walleye (*Stizostedion vitreum*) Eggs in a Northern Ontario River," *Ecoscience* 1, no. 3 (1994): 215–22.

8. G. G. Sass, J. F. Kitchell, S. R. Carpenter, T. R. Hrabik, A. E. Marburg, and M. G. Turner, "Fish Community and Food Web Responses to a Whole-Lake Removal of Coarse Woody Habitat," *Fisheries* 31, no. 7 (2006): 321–30.

9. Lynn Markham, "Shoreland Development Density and Impervious Surfaces" (University of Wisconsin–Stevens Point, Center for Land Use Education, 2003), 14.

Chapter 7. Zoning and Its Discontents

1. Frank Zufall, "State Regulations Could Override County Lakeshore Zoning if Approved," *Sawyer County Record*, June 26, 2015.

2. *Protecting Michigan's Inland Lakes: A Guide for Local Governments* (Paw Paw, MI: Van Buren Conservation District, n.d.), 3, accessed January 2022, https://www.michigan.gov/documents/deq/wrd-nps-inland-lakes-guide_634823_7.pdf.

3. Grenetta Thomassey, *Charlevoix County: Local Ordinance Gaps Analysis* (Petosky, MI: Tip of the Mitt Watershed Council, 2011), 23.

4. Thomassey, *Charlevoix County: Local Ordinance Gaps Analysis*, 102.

5. Grenetta Thomassey, *Antrim County Local Ordinance Gaps Analysis* (Petosky, MI: Tip of the Mitt Watershed Council, 2011), 52.

6. Grenetta Thomassey, *Cheboygan County Local Ordinance Gaps Analysis* (Petosky, MI: Tip of the Mitt Watershed Council, 2011), 25

Chapter 8. Defective Septic Systems

1. Marc P. Verhougstraete, Sherry L. Martin, Anthony D. Kendall, David W. Hyndman, and Joan B. Rose, "Linking Fecal Bacteria in Rivers to Landscape, Geochemical, and Hydrologic Factors and Sources at the Basin Scale," *Proceedings of the National Academy of Sciences* 112, no. 33 (August 18, 2015): 10419–24.

2. "Septic Tanks Aren't Keeping Poo Out of Rivers and Lakes," MSU Today, Michigan State University, August 3, 2015, https://msutoday.msu.edu/news/2015/septic-tanks-arent-keeping-poo-out-of-rivers-and-lakes.

3. "Septic System Improvement Estimator," Onsite Sewage Treatment Program, University of Minnesota, accessed, April 2020, https://septic.umn.edu/ssts-professionals/forms-worksheets#SSIE.

4. "The Septic Question," project report, Health Department of Northwest Michigan and Tip of the Mitt Watershed Council, 2016, 12, https://www.managingwaterforhealth.org/wp-content/uploads/the_septic_question_report-final-web2.pdf.

5. Brad Neumann, "Got Water? Time of Sale Septic Inspections Can Protect Public Water Quality, Part 1," Michigan State University Extension, October 6, 2015, https://www.canr.msu.edu/news/got_water_time_of_sale_septic_inspections_can_protect_water_quality_part_1.

6. Ron Seely, "Human Waste Pollutes Some Wisconsin Drinking Water," Wisconsin Watch, Wisconsin Center for Investigative Journalism, May 2, 2016, https://wisconsinwatch.org/2016/05/human-waste-pollutes-some-wisconsin-drinking-water/.

7. "Why Do We Need This Program?," Time of Sale Program, Washtenaw County, Michigan, accessed December 16, 2021, https://www.washtenaw.org/1727/Time-of-Sale-Program-TOS.

8. Wisconsin Statutes 145.245(4).

9. "The Septic Question," 14.

10. A Secchi disk is a scientific tool used to measure how deep light penetrates into a body of water. It consists of a disc with alternating white and black quadrants that is lowered into the water by a rope. It is a standard measure of water clarity.

Chapter 9. In the Wake

1. Scott Brown, "A Steadily Increasing Number of Wake Boats Operating on Michigan's Inland Waters Are Rendering Damage to Shallow Water Habitat and to

Developed and Natural Shorelines," Michigan Waterfront Alliance, August 24, 2020, https://michiganwaterfrontalliance.com/2020/08/24/a-steadily-increasing-num ber-of-wake-boats-operating-on-michigans-inland-waters-are-rendering-damage -to-shallow-water-habitat-and-to-developed-and-natural-shorelines/.

2. Clifford H. Bloom; "The Killer Bees Appear to Be Winning: An Update on Wake Boats," Michigan Lakes and Streams Association, October 9, 2017, https:// mymlsa.org/the-killer-bees-appear-to-be-winning-an-update-on-wake-boats.

3. Rachel Bachman, "Far from the Ocean, These Surfers Don't Wait for Waves," *Wall Street Journal*, May 17, 2015.

4. "U.S. Boat Sales Reached 13-Year High in 2020," National Marine Manufac- turers Association, January 6, 2021, https://www.nmma.org/press/article/23527.

5. "Wave Type Comparison," Guinn Partners, accessed June 2020, https://www .guinnpartners.com/boat-testing/.

6. This information is drawn from Rich Armstrong, "The Science Behind How Wake Boats Create Surfable Waves," BoatUS, April 2018, https://boatus.com/expert -advice/expert-advice-archive/2018/april/wake-boats.

7. "The Biggest Wakesurfing Wave on Earth," Pavati Wake Boat, accessed June 2020, https://pavati.com/the-biggest-wake-surfing-wave-on-earth.

8. "Big Wave Surfing," Gigawave, accessed June 2020, https://www.ridegiga wave.com.

9. "ZS232," Supreme Towboats, accessed January 2020, https://supremetow boats.com/boats/zs232/.

10. "WSIA Wae Energy Study Summary," Water Sports Industry Association, accessed June 2020, https://www.wsia.net/wp-content/uploads/2019/05/WSIA -Wave_Energy_Study_Summary.pdf.

11. "Own Your Wake," Minnesota Department of Natural Resources, accessed June 2020, https://www.dnr.state.mn.us/safety/boatwater/own-your-wake.html.

12. "Take Action," Midwest Wakesurf Association, accessed January 2022, https://midwestwakesurfing.com/take-action/.

13. SafeWakes for Minnesota Lakes, accessed August 2020, http://www.safe wakes.org.

14. "Fishermen Jump for Water Before Boat Collision on Jewett Lake," *Perham Focus*, July 6, 2020.

15. Sara Mercier-Blais and Yves Prairie, "Project Evaluation of the Impact of Waves Created by Wake Boats on the Shores of Lake Memphremagog and Lover- ing," Université du Québec à Montréal, June 2014, 22, http://gencourt.state.nh.us/ statstudcomm/committees/1434/documents/Impact%20°f%20Waves%20Cre ated%20by%20Wake%20Boats-%20Canada.pdf.

16. Sebastien Raymond, "Impact of the Navigation in Milieu Lacustre," Univer- sité Laval, November 25, 2015, https://www.documentcloud.org/documents/6801 170-S%C3%A9Bastien-2015-English-U-of-Laval-1.html.

17. Jeffrey Marr, Andrew Riesgraf, William Herb, Matthew Lueker, Jessica Koza- rek, and Kimberly Hill, "A Field Study of Maximum Wave Height, Total Wave Energy,

and Maximum Wave Power Produced by Four Recreational Boats on a Freshwater Lake," St. Anthony Falls Laboratory, University of Minnesota, Department of Civil, Environmental, and Geo-Engineering, February 2022, https://conservancy.umn.edu/handle/11299/226190.

18. JoAnn Syverson, "Counterpoint: A 200-Foot Wake Restriction? Not Enough," *Star-Tribune*, March 11, 2020.

19. Prohibited Operation to Enhance Wake, City of Mequon, WI, Municipal Code, Sec. 90-17.

20. Motorboat Wake Protection Area Ordinance No. 2018-10-08, Town of Bass Lake, WI.

21. An Act Relating to Regulating the Use of Wake Boats on State Waters, Vermont S.69 (introduced February 1, 2019).

22. "NMMA Sounds the Alarm on Vermont Legislation Banning Wake Surfing," National Marine Manufacturers Association, February 8, 2019, https://www.nmma.org/press/article/22495.

23. "New Wake Boat Study Confirms Industry-Backed Positions," Boating Industry, July 28, 2020, https://boatingindustry.com/news/2020/07/28/new-wake-boat-study-confirms-industry-backed-positions/.

24. Corey Buchanan, "Riverkeeper Group Requests Ban on Willamette Wake Boats," *West Linn Tidings*, November 7, 2019.

25. "Idaho County Drops Proposed Wake Boat Restriction After Strong Public Opposition," National Marine Manufacturers Association, May 10, 2019, https://www.nmma.org/press/article/22664.

26. Citizens for Sharing Lake Minnetonka, accessed September 2021, https://sharelakeminnetonka.com/.

Chapter 10. Stealth Invaders

1. "Cornell Student Finds Invasive Water Flea in Oneida Lake," *Cornell Chronicle*, October 9, 2019, https://news.cornell.edu/stories/2019/10/cornell-student-finds-invasive-water-flea-oneida-lake.

2. "Spiny Waterflea," Vermont Invasives, accessed July 2020, https://vtinvasives.org/invasive/spiny-waterflea.

3. "Spiny Water Flea (*Bythotrephes longimanus*)," Vander Zanden Lab, Center for Limnology, University of Wisconsin–Madison, accessed July 2020, https://www.jakevzlab.net/spiny-water-flea.html.

4. Brett Kelly and Ben Martin, "Food Web Ecology and Science Communication with Ben Martin," *The Fisheries Podcast*, August 2, 2020, https://www.stitcher.com/podcast/the-fisheries-podcast/e/76628730?autoplay=true.

5. Gretchen J. A. Hansen, Tyler D. Ahrenstorff, Bethany J. Bethke, Joshua D. Dumke, Jodie Hirsch, Katya E. Kovalenko, Jaime F. LeDuc, Ryan P. Maki, Heidi M. Rantala, and Tyler Wagner, "Walleye Growth Declines Following Zebra Mussel and *Bythotrephes* Invasion," *Biological Invasions* 22, no. 4 (April 2020): 1481–95, https://doi.org/10.1007/s10530-020-02198-5.

6. "Big Consequences of Small Invaders: Adaptable Spine Helps Spiny Water-flea Fend Off Predators," eForum, Great Lakes Fishery Commission newsletter, accessed August 2020, http://www.glfc.org/eforum/article9.html/.

7. "What Lies Beneath: Sudden Invasion of a Wisconsin Lake Wasn't So Sudden After All," Water Blogged, Center for Limnology, University of Wisconsin–Madison, March 22, 2017, https://blog.limnology.wisc.edu/2017/03/22/what-lies-beneath-sudden-invasion-of-a-wisconsin-lake-wasnt-so-sudden-after-all/; "Cascading Impact of Spiny Water Flea on Ecosystem Services in Lake Mendota," Jake R. Walsh personal website, accessed September 2020, https://jakerwalsh.wixsite.com/jakerwalsh/thesis-work.

8. "Field Samples: Spiny Water Fleas, Lake Mendota, And Green Water," Water Blogged, Center for Limnology, University of Wisconsin–Madison, January 28, 2015, https://blog.limnology.wisc.edu/2015/01/28/field-samples-jake-walsh/.

9. W. Charles Kerfoot, Foad Yousef, Martin M. Hobmeier, Ryan P. Maki, S. Taylor Jarnagin, James H. Churchill, "Temperature, Recreational Fishing and Diapause Egg Connections: Dispersal of Spiny Water Fleas (*Bythotrephes longimanus*), *Biological Invasions* 13, no. 11 (November 2011): 2513–31, https://doi.org/10.1007/s10530-011-0078-8.

10. John Myers, "Blame Us, Not Ducks, for Spreading Spiny Water Fleas," *Duluth News Tribune*, August 29, 2020.

11. Donn K. Branstrator, Joshua D. Dumke, Valerie J. Brady, and Holly A. Wellard Kelly, "Lines Snag Spines! A Field Test of Recreational Angling Gear Ensnarement of *Bythotrephes*," *Lake and Reservoir Management* 37, no. 4 (2021): 391–405, https://doi.org/10.1080/10402381.2021.1941447.

12. Myers, "Blame Us, Not Ducks."

13. Kerfoot et al., "Temperature, Recreational Fishing and Diapause Egg Connections."

14. "Zebra Mussel (*Dreissena polymorpha*)," Minnesota Department of Natural Resources, accessed December 11, 2021, https://www.dnr.state.mn.us/invasives/aquaticanimals/zebramussel/index.html.

15. AIS Smart Prevention Tool 2.0, Center for Limnology, University of Wisconsin–Madison, accessed August 2020, https://uwlimnology.shinyapps.io/AISSmartPrevention2.

16. "Zebra Mussels Threaten Inland Waters: An Overview," Minnesota Sea Grant, March 30, 2017, http://www.seagrant.umn.edu/ais/zebramussels_threaten.

17. Steven Verburg, "Zebra Mussels May Worsen Lake Health Hazards Caused by Farm Pollution," *Wisconsin State Journal*, June 24, 2017.

18. Dan Gunderson, "Measuring the Impact of Invasives in Minnesota Lakes Is Complicated, Inconclusive," Minnesota Public Radio News, August 27, 2018, https://www.mprnews.org/story/2018/08/27/measuring-impact-of-aquatic-invasive-species-in-minnesota-complicated-inconclusive.

19. Tom Jones, "The Problem of Zeebs," *Hooked on Mille Lacs Lake* 4 (February 2015): 5–6.

20. Hansen et al., "Walleye Growth Declines."

21. Gretchen J. A. Hansen, Luke A. Winslow, Jordan S. Read, Melissa Treml, Patrick J. Schmalz, and Stephen R. Carpenter, "Water Clarity and Temperature Effects on Walleye Safe Harvest: An Empirical Test of the Safe Operating Space Concept," *Ecosphere* 10, no. 5 (May 2019), e02737, https://doi.org/10.1002/ecs2.2737.

22. Gunderson, "Measuring the Impact of Invasives."

23. "Project Successfully Removes Invasive Quagga Mussels Near Sleeping Bear Dunes in Lake Michigan," Great Lakes Commission, December 8, 2020, https://www.glc.org/news/ghr-120820.

24. "MAISRC Subproject 36: RNA-Interference Screens for Zebra Mussel Bio-control Target Genes," Environment and Natural Resources Trust Fund (ENRTF) M.L. 2019 Minnesota Aquatic Invasive Species Research Center Subproject Work Plan, August 28, 2020, https://mckenzielakes.com/wp-content/uploads/sites/73/2020/09/ZEBRA-MUSSEL-U-OF-M-RESEARCH-GRANT.pdf.

25. "Exciting Findings in Recent Zebra Mussel Research," Minnesota Aquatic Invasive Species Research Center newsletter, April 2020, https://maisrc.umn.edu/newsletter-april2020.

Chapter 11. Preventing the Spread

1. Quoted in Peter Jurich, "Time to Reconsider the Language of the Past," in "Embracing the Benefits of Aquatic Plants," special insert, *Wisconsin Natural Resources* (Fall 2020): 8.

2. Alison Mikulyuk, Ellen Kujawa, Michelle E. Nault, Scott Van Egeren, Kelly I. Wagner, Martha Barton, Jennifer Hauxwell, M. Jake Vander Zanden, and Peter G. Kevan, "Is the Cure Worse than the Disease? Comparing the Ecological Effects of an Invasive Aquatic Plant and the Herbicide Treatments Used to Control It," *FACETS* 5, no. 1 (2020): 353–66, https://doi.org/10.1139/facets-2020-0002.

3. Paul Radomski and Donna Perleberg, "Avoiding the Invasive Trap: Policies for Aquatic Non-Indigenous Plant Management," *Environmental Values* 28, no. 2 (2019): 211–32, https://doi.org/10.3197/096327119X15515267418539.

4. Minnesota Department of Natural Resources, *Invasive Species 2019 Annual Report*, https://files.dnr.state.mn.us/natural_resources/invasives/2019-invasive-species-annual-report.pdf.

5. Dave Orrick, "Zebra Mussels' Best Friend: Wakeboard Boats, New U Study Finds," *Pioneer Press*, January 22, 2019.

6. State of New Hampshire, *Final Report of the Commission to Study Wake Boats*, June 30, 2020, https://nhlakes.org/wp-content/uploads/Commission-to-Study-Wake-Boats-Final-Report-1.pdf.

7. "Mussel Mast'R," Wake Worx, accessed October 2020, https://wake-worx.com/shop/mussel-mastr/.

8. Michigan Invasive Species Program, "2019 Annual Report," 6, https://www.michigan.gov/documents/invasives/2019_Invasive_Species_annual_report_optimized_694011_7.pdf.

9. Michigan Invasive Species, accessed January 2020, https://www.michigan .gov/invasives/0,5664,7-324-103844_68072---,00.html.

10. Minnesota Department of Natural Resources, *Invasive Species 2019 Annual Report.*

11. Otter Tail County, Minnesota, Aquatic Invasive Species Prevention Program Summary, 2018, p. 4.

12. Eric Lindberg and Environmental Sentry Protection, "AIS Prevention through a Boat Tagging Model," 2015.

Chapter 12. Changing Climate, Changing Lakes

1. Courte Oreilles Lakes Association, "Report on 2016 Fish Kill, Lac Courte Oreilles, Sawyer County," October 6, 2016, 5.

2. "World Water Day 2020: Water and Climate Change," North American Lake Management Society, March 2020, https://www.nalms.org/world-water-day-2020.

3. Quoted in "Climate Change: How Do We Know?," National Aeronautics and Space Administration (NASA), accessed March 2022, https://climate.nasa.gov/ evidence/.

4. "Climate Change: How Do We Know?"

5. "Climate Change: Global Temperature," NOAA Climate.gov, accessed January 2021, https://www.climate.gov/news-features/understanding-climate/climate -change-global-temperature.

6. "Climate Change: How Do We Know?"

7. Steve Vavrus, "Climate Change and Extreme Weather in Wisconsin" (presentation, Wisconsin Lakes and Rivers Convention Online Learning Event, April 2, 2020), https://youtu.be/ePhFTQ3TuU8.

8. Kyuhyun Byun and Alan F. Hamlet, Projected Changes in Future Climate over the Midwest and Great Lakes Region Using Downscaled CMIP5 Ensembles," *International Journal of Climatology* 38, no. S1 (April 2018): e531–e553, https://doi .org/10.1002/joc.5388.

9. Kyuhyun Byun, Chun-Mei Chiu, and Alan F. Hamlet, "Effects of 21st Century Climate Change on Seasonal Flow Regimes and Hydrologic Extremes over the Midwest and Great Lakes Region of the US," *Science of the Total Environment* 650, pt. 1 (February 2019): 1261–77, https://doi.org/10.1016/j.scitotenv.2018.09.063.

10. "Ice Cover," Climate Wisconsin, accessed December 2020, https://climate wisconsin.org/story/ice-cover.html.

11. "Climate Change Indicators: Lake Ice," United States Environmental Protection Agency, accessed January 2021, https://www.epa.gov/climate-indicators/climate -change-indicators-lake-ice.

12. "Ice Cover."

13. "Climate Change Impacts on Lakes," North American Lake Management Society, February 5, 2015, https://www.nalms.org/nalms-position-papers/climate -change-impacts-on-lakes/.

14. "Blame it on the Rain: Study Ties Nutrient Loading in Lakes to Extreme Precipitation Events." Water Blogged, Center for Limnology, University of Wisconsin–Madison, January 11, 2018, https://blog.limnology.wisc.edu/2018/01/11/blame-it-on-the-rain-study-ties-phosphorus-loading-in-lakes-to-extreme-precipitation-events/.

15. Lisa Borre, "Climate Change Already Having Profound Impacts on Lakes in Europe," *National Geographic*, July 21, 2014.

16. Grace M. Wilkinson, Jonathan A. Walter, Cal D. Buelo, and Michael L. Pace, "No Evidence of Widespread Algal Bloom Intensification in Hundreds of Lakes," *Frontiers in Ecology and the Environment* 20, no. 1 (February 2022): 16–21, https://doi.org/10.1002/fee.2421.

17. Ke Xiao, Timothy J. Griffis, John M. Baker, Paul V. Bolstad, Matt D. Erickson, Xuhui Lee, Jeffrey D. Wood, Cheng Hu, and John L. Nieber, "Evaporation from a Temperate Closed-Basin Lake and Its Impact on Present, Past, and Future Water Level," *Journal of Hydrology* 561 (June 2018): 59–75, https://doi.org/10.1016/j.jhydrol.2018.03.059.

18. Zachary Hanson, "Integrated Surface Water and Groundwater Modeling of Inland Lakes and Wetlands in a Changing Climate" (PhD diss., University of Notre Dame, 2019).

19. "Climate Change Effects on Michigan's Fisheries," Michigan Department of Natural Resources, accessed January 2021, https://www.michigan.gov/dnr/0,4570,7-350--403361--,00.html.

20. Elizabeth Weise, "Global Warming Could Mean Fewer Fish for Sport Fishing, More Die-Offs Across US," *USA Today*, July 9, 2019, https://www.usatoday.com/story/news/nation/2019/07/09/global-warming-killing-fish-hurting-sport fishing-industry/1675771001.

21. Gretchen J. A. Hansen, Jordan S. Read, Jonathan F. Hansen, and Luke A. Winslow, "Projected Shifts in Fish Species Dominance in Wisconsin Lakes Under Climate Change, *Global Change Biology* 23, no. 4 (April 2017): 1463–76, https://doi.org/10.1111/gcb.13462.

22. "Warmer Waters from Climate Change Could Impact Sport Fish Communities in Midwestern Lakes," United States Geological Survey, September 12, 2016, https://www.usgs.gov/news/warmer-waters-climate-change-could-impact-sport-fish-communities-midwestern-lakes.

23. "As Lakes Grow Warmer, the Race Is on to Save Minnesota's Cold-Water Fish," *Star Tribune*, July 27, 2019.

24. Chris O'Brien, "Refuge for Tullibees: Protecting Private Forests Could Help Save Minnesota's Tullibees, Important Forage Fish for Many Game Fish Species," *Minnesota Conservation Volunteer*, May–June 2016, https://www.dnr.state.mn.us/mcv magazine/issues/2016/may-jun/refuge-lakes.html.

25. John Lyons, Jeff Kampa, Tim Parks, and Greg Sass, "The Whitefishes of Wisconsin's Inland Lakes: The 2011–2014 Wisconsin Department of Natural Resources Cisco and Lake Whitefish Survey," Fisheries and Aquatic Research Section, Wisconsin DNR, February 2015, https://nelson.wisc.edu/wp-content/uploads/lyons.pdf.

26. "Cisco (Lake Herring)," Michigan Sea Grant, accessed January 2021, https://www.michiganseagrant.org/topics/ecosystems-and-habitats/native-species-and-biodiversity/cisco-lake-herring/.

27. Quoted in O'Brien, "Refuge for Tullibees."

28. Dan Kraker, "As State Warms, a Few Spots Keep Their Cool," Minnesota Public Radio News, February 3, 2015, https://www.mprnews.org/story/2015/02/03/climate-change-coldspots.

29. Quoted in Borre, "Climate Change Already Having Profound Impacts on Lakes in Europe."

Chapter 13. Loons under Stress

1. Kristin Bianchini, Douglas C. Tozer, Robert Alvo, Satyendra P. Bhavsar, Mark L. Mallory, "Drivers of Declines in Common Loon (*Gavia immer*) Productivity in Ontario, Canada," *Science of the Total Environment* 738 (2020): e139724, https://doi.org/10.1016/j.scitotenv.2020.139724.

2. Walter H Piper, Jason Grear, Brian Hoover, Elaina Lomery, Linda M Grenzer, "Plunging Floater Survival Causes Cryptic Population Decline in the Common Loon," *Ornithological Applications* 122, no. 4 (2020): duaa044, https://doi.org/10.1093/condor/duaa044.

3. See "2020 Wisconsin Loon Population Survey Results," Turtle-Flambeau Flowage & Trude Lake Property Owners' Assoc., https://tfftl.org/2020-wisconsin-loon-population-survey-results/.

4. "Common Loon," Minnesota Department of Natural Resources, accessed March 2022, https://www.dnr.state.mn.us/birds/commonloon.html.

5. "Eggs and Predators," Loon Preservation Committee (New Hampshire), June 7, 2017, https://loon.org/looncam-blogs/loon-cam-2017/eggs-and-predators.

6. "State's Loon Population on the Rise," *The Chronotype*, May 23, 2016.

7. Minnesota Loon Monitoring Program, "2020 Annual Report," Minnesota Department of Natural Resources, https://files.dnr.state.mn.us/eco/nongame/projects/mlmp-2020.pdf.

8. "FAQ's about Loons & Lead Poisoning," Loon Preservation Committee (New Hampshire), accessed December 2020, https://loon.org/loons-and-lead/faqs.

9. "Get the Lead Out: Lead Poisoning of a Loon," Minnesota Department of Natural Resources, accessed December 2020, https://www.dnr.state.mn.us/eco/nongame/projects/leadout.html.

10. "Protecting Loons from Lead," Loon Preservation Committee (New Hampshire), accessed December 2020, https://loon.org/loons-and-lead.

11. "Maine Loom Mortality, 1987–2012," Maine Audubon, February 2013, https://www.maineaudubon.org/wp-content/uploads/2017/05/REPORT-Effects-of-Lead-Fishing-Tackle-on-Loons-in-Maine.pdf.

12. Tiffany J. Grade, Mark A. Pokras, Eric M. Laflamme, and Harry S. Vogel, "Population-Level Effects of Lead Fishing Tackle on Common Loons," *Journal of Wildlife Management* 82, no. 1 (January 2018): 155–64, https://doi.org/10.1002/jwg.21348.

13. "FAQ's about Loons & Lead Poisoning."

14. Howard Meyerson, "Lead and Loons," *Outdoor Journal*, July 8, 2015, https://howardmeyerson.com/2015/07/08/lead-and-loons.

15. "Lead-Free Program for MN Loons Gets Green Light," Minnesota Public Radio News, February 19, 2020, https://www.mprnews.org/story/2020/02/19/leadfree-program-for-mn-loons-gets-green-light.

16. "The Practical Impacts of Banning Lead Sinkers for Fishing," American Sportfishing Association, June 2011, https://asafishing.org/uploads/Lead_in_Fishing_Tackle.pdf.

17. "Why Are We Still Killing Loons with Lead Sinkers?," Loon Project, September 9, 2020, https://loonproject.org/2020/09/09/why-are-we-still-killing-loons-with-lead-sinkers.

18. "A Human Family Saves a Loon Family," Loon Project, August 24, 2020, https://loonproject.org/2020/08/24/a-human-family-saves-a-loon-family.

19. *Loons, Lead, and Line Don't Mix* (Saranac Lake, NY: Adirondack Center for Loon Conservation, n.d.).

20. "More Boaters Mean More Threats to Loons on Minn. Lakes," Minnesota Public Radio News, July 28, 2020, https://www.mprnews.org/story/2020/07/28/more-boaters-mean-more-threats-to-loons.

21. "More Boaters Mean More Threats to Loons."

22. *Survival by Degrees: 389 Bird Species on the Brink* (New York: National Audubon Society, 2019).

23. Michael W. Meyer, John F. Walker, Kevin P. Kenow, Paul W. Rasmussen, Paul J. Garrison, Paul C. Hanson, and Randall J. Hunt, "Executive Summary," in *Potential Effects of Climate Change on Inland Glacial Lakes and Implications for Lake-Dependent Biota in Wisconsin* (Madison, WI: Environmental and Economic Research and Development Program, Focus on Energy, April 2013), viii.

24. "A More Balanced View," Loon Project, March 21, 2020, https://loonproject.org/2020/03/21/a-more-balanced-view.

Chapter 14. Pressure Rising

1. "Six Ways that Short-Term Vacation Rentals Are Impacting Communities," Granicus Blog, accessed February 2021, https://granicus.com/blog/six-ways-that-short-term-vacation-rentals-are-impacting-communities/.

2. "Short-Term Rental Controversy Continues," Rosi & Gardner, P.C., March 16, 2020, https://rosigardner.com/short-term-rentals-controversy-continues/.

3. Matt Mikus, "Real Estate Market Keeps Buzzing into Slowdown Season," Minnesota Public Radio News, January 4, 2021, https://www.mprnews.org/story/2021/01/04/real-estate-market-keeps-buzzing-into-slowdown-season.

4. Jim Skibo, "Northwoods Real Estate Boom During Pandemic," WXPR Radio, September 11, 2020, https://www.wxpr.org/post/northwoods-real-estate-boom-during-pandemic#stream/0.

5. Zholdas Orisbayev, "Northern Michigan Real Estate Booms as Builders Struggle to Meet Demand," *Detroit News*, December 19, 2020.

6. "U.S. Boat Sales Reached 13-Year High in 2020, Recreational Boating Boom to Continue through 2021," National Marine Manufacturers Association, January 6, 2021, https://www.nmma.org/press/article/23527.

7. Dan Gunderson, "Another COVID-19 Ripple Effect: Crowded Lakes Cause Conflict," Minnesota Public Radio News, July 23, 2020, https://www.mprnews.org/story/2020/07/23/another-covid19-ripple-effect-crowded-lakes-cause-conflict.

8. "Gunderson, "Another COVID-19 Ripple Effect."

9. Adam Hinterhuer, "Study Says 'Hidden Overharvest' from Fishing Plays a Role in Wisconsin Walleye Declines," University of Wisconsin News, November 18, 2019, https://news.wisc.edu/study-says-hidden-overharvest-from-fishing-plays-a-role-in-wisconsin-walleye-declines.

10. Erick Elgin, "Salt Runoff Can Impair Lakes," Michigan State University Extension, February 22, 2018, https://www.canr.msu.edu/news/salt_runoff_can_impair_lakes.

11. Hilary A. Dugan and Linnea A. Rock, "The Slow and Steady Salinization of Sparkling Lake, Wisconsin," *Limnology and Oceanography Letters*, April 5, 2021, https://doi.org/10.1002/lol2.10191.

12. Hilary Dugan and Bill Hintz, "Salty Streams and Formerly Freshwater Lakes: An Ecosystem Perspective," Wisconsin Salt Awareness Week presentation, January 11, 2021, https://www.youtube.com/watch?v=0z9K35HdYdg/.

13. Emily Sohn, "Hold the Salt," *Minnesota Conservation Volunteer*, January–February 2020.

Chapter 15. Ways Forward

1. "The National Lakes Assessment (NLA) 2012," United States Environmental Protection Agency, https://www.epa.gov/sites/default/files/2016-12/documents/nla_fact_sheet_dec_7_2016.pdf.

2. "The National Lakes Assessment (NLA) 2012."

3. "National Lakes Assessment 2012 Key Findings," United States Environmental Protection Agency, photo 9, https://19january2017snapshot.epa.gov/national-aquatic-resource-surveys/national-lakes-assessment-2012-key-findings_.html.

4. Katie Hien and Ali Mikulyuk, "A Snapshot of Lake Health Across Wisconsin" (PowerPoint presentation, 2018 Lakes Partnership Convention, Stevens Point, WI, April 19, 2018), https://www.uwsp.edu/cnr-ap/UWEXLakes/Documents/programs/convention/2018/TH-Session1/KatieHeinAliMikulyuk_ASnapshotofLakeHealthAcrossWisconsin.pdf.

5. Erick Elgin, "The State of Michigan's Inland Lake Shorelines," Michigan State University Extension, November 17, 2017, https://www.canr.msu.edu/news/the_state_of_michigans_inland_lake_shorelines.

6. Jesse Anderson, Allison Gamble, and Lee Engel, *National Lakes Assessment 2017* (St. Paul, MN: Minnesota Pollution Control Agency, November 2020), https://www.pca.state.mn.us/sites/default/files/wq-nlap1-17.pdf.

7. "She Walked the Talk: Farewell to Water Warrior Grandmother Josephine Mandamin," Water Docs, February 22, 2019, https://www.waterdocs.ca/news/2019/2/22/she-walked-the-talk-farewell-to-water-warrior-grandmother-josephine-man damin.

8. "News from the Lac du Flambeau Tribe," February 10, 2017, https://myemail.constantcontact.com/News-from-the-Lac-du-Flambeau-Tribe.html?soid=11230577 92622&aid=vU8T7hvkNrU.

9. "About Us," Lac du Flambeau Tribe, accessed March 2021, https://www.ldf tribe.com/pages/2/about-us/.

10. "News from the Lac du Flambeau Tribe."

11. "News from the Lac du Flambeau Tribe."

12. Paul Radomski and Kristof Van Assche, *Lakeshore Living: Designing Lake Places and Communities in the Footprints of Environmental Writers* (East Lansing: Michigan State University Press, 2014).

13. "Bemidji State University Study Reveals Water Clarity Is Most Important Factor in Determining Lakeshore Property Values," *Brainerd Dispatch*, May 31, 2003, https://www.brainerddispatch.com/sports/3454004-bemidji-state-university-study -reveals-water-clarity-most-important-factor.

14. "Tainter Lake & Lake Menomin: The Impact of Diminishing Water Quality on Value," Fiscal and Economic Research Center, University of Wisconsin–Whitewater, accessed March 2021, http://www.uww.edu/Documents/colleges/cobe/ferc/ TainterLakes.pdf.

15. Sarah Nicholls and John Crompton, "A Comprehensive Review of the Evidence of the Impact of Surface Water Quality on Property Values," *Sustainability* 10, no. 2 (February 2018), https://doi.org/10.3390/su10020500.

16. Eric James Horsch, "Quantifying the Economic Effects of Invasive Species: A Non-Market Valuation Analysis of Eurasian Water Milfoil" (master's thesis, University of Wisconsin–Madison, 2008), https://lter.limnology.wisc.edu/sites/default/ files/Horsch_Thesis_Final.pdf.

17. Thomas Kemp and David Wolf, "Relationship Between Lake Water Clarity and Residential Housing Value in Vilas and Oneida Counties" (PowerPoint presentation, Wisconsin Lakes Partnership Convention, University of Wisconsin–Eau Claire, April 11, 2019), https://www.uwsp.edu/cnr-ap/UWEXLakes/Documents/ programs/convention/2019/TH-Session3/ThomasKemp_LakeQuality-Residential Living.pdf.

18. David Noel, "Economic Value of Lakes and Rivers in Oneida County," Oneida County Land and Water Conservation Department and University of Wisconsin Extension, 2019, http://www.oclra.org/uploads/7/4/3/4/74342595/flier_-_econ_value _of_oc_lakes_2019.pdf.

19. "Rate Your Shoreland," Michigan Shoreland Stewards, accessed March 2021, https://www.mishorelandstewards.org/rate.asp.

20. "CLMP Documents," Michigan Clean Water Corps, accessed March 2021, https://micorps.net/lake-monitoring/clmp-documents/.

21. "Score Your Shore: A Citizen Shoreline Description Survey," Minnesota Department of Natural Resources, accessed March 2021, https://www.dnr.state.mn.us/scoreyourshore/index.html.

22. "Shoreland Evaluation Tool," Healthy Lakes and Rivers, accessed August 2021, https://survey.healthylakeswi.com/.

23. Carrie Rasmussen, "Yawkey Forest Reserve Is One of the Largest Donations of Conservation Land in Wisconsin," Northwoods Land Trust, April 29, 2019, https://northwoodslandtrust.org/yawkey-forest-reserve-is-one-of-the-largest-donations-of-conservation-land-in-wisconsin/.

24. Paul Radomski and Kristin Carlson, "Prioritizing Lakes for Conservation in Lake-Rich Areas," *Lake and Reservoir Management* 34, no. 4 (2018): 401–416, https://doi.org/10.1080/10402381.2018.1471110.

Interviews

Chapter 1

Tom Rulseh, McDonald Lake property owner, Vilas County, WI, November 11, 2019.

Chapter 2

Denny Thompson, Birch Lake property owner, Oneida County, WI, December 2019.

Chapter 3

Sandra Ebben, First Weber Realtors, Rhinelander, WI, January 17, 2020.
Kyle Zastrow, real estate appraisal specialist, Rhinelander, WI, February 25, 2020.
Jeff Forester, executive director, Minnesota Lakes and Rivers Advocates, March 19, 2020.

Chapter 5

BreAnne Grabill, manager of Northern Regional Lakes for PLM Lake and Land Management Corporation, May 2020.
Jean Roach, Pelican Lake Property Owners Association, Oneida County, WI, March 11, 2020.
Dave Hardt, Pelican Lake Property Owners Association, Oneida County, WI, March 24, 2020.
Bob Martini, Oneida County Lakes and Rivers Association and Wisconsin's Green Fire, March 2020.
Jennifer Buchanan, associate director, Tip of the Mitt Watershed Council, MI, April 13, 2020.

Chapter 6

Patrick Goggin, lake specialist, University of Wisconsin Extension, March 4, 2020.
Heidi Shaffer, soil erosion officer, Antrim Conservation District, MI, April 16, 2020.

Julia Kirkwood, Michigan Department of Environment, Great Lakes, and Energy, May 15, 2020.

Paul Radomski, research scientist, Minnesota Department of Natural Resources, April 23, 2020.

Chapter 7

John Richter, Plum Lake Association, Vilas County, WI, May 18, 2020.

Daniel Petrik, land use specialist, Minnesota Department of Natural Resources, May 20, 2020.

Michael Engleson, executive director, Wisconsin Lakes, May 2020.

Eric Calabro, inland lakes analyst, Michigan Department of Environment, Great Lakes and Energy, June 9, 2020.

Bob Martini, Oneida County Lakes and Rivers Association and Wisconsin's Green Fire, March 2020.

Jay Kozlowski, zoning and conservation administrator, Sawyer County, WI, May 27, 2020

Dawn Schmidt, zoning administrator, Vilas County, WI, June 10, 2020.

Chapter 8

Rollie Mann, former administrator, Otter Tail Water Management District, MN, August 26, 2020.

Sara Heger, Onsite Sewage Treatment Program, University of Minnesota Water Resources Center, June 4, 2020.

Jim Anderson, former director, University of Minnesota Water Resources Center, June 2020.

Eric Wellauer, county sanitarian, Sawyer County, WI, August 21, 2020.

Merton Maki, soil tester and former county sanitarian, Sawyer County, WI, August 2020.

Chapter 9

Michael Engleson, executive director, Wisconsin Lakes, September 9, 2020.

Jeff Forester, executive director, Minnesota Lakes and Rivers Advocates, September 28, 2020.

Larry Meddock, chairman, Water Sports Industry Association, September 21, 2020.

Dan Butkus, Squash Lake Protection and Rehabilitation District, Oneida County, WI, September 1, 2020.

Timothy Tyre, North Lake Management District, Waukesha County, WI, September 8 and 17, 2020.

Chuck Becker, SafeWakes for Minnesota Lakes, September 16, 2020.

Chris Bischoff, waterways access and government relations specialist, Water Sports Industry Association, September 21, 2020.

Dave Maturen, president, Michigan Lakes and Streams Association, September 22, 2020.

Chapter 10

Sandra Swanson, McKenzie Lakes Association, Burnett and Washburn counties, WI, November 3, 2020.

Thomas Boisvert, aquatic invasive species coordinator, Burnett County, WI (now conservation program manager, Lincoln County, WI), November 16, 2020.

Ben Martin, food web ecologist, Center for Limnology, University of Wisconsin–Madison, October 19, 2020.

Tom Heinrich, Mille Lacs Area fisheries supervisor, Minnesota Department of Natural Resources, November 2020.

Tom Jones, regional treaty coordinator for fisheries and wildlife, Minnesota Department of Natural Resources, October 30, 2020.

Caroline Keson, monitoring program coordinator, Tip of the Mitt Watershed Council, MI, October 2020.

Chapter 11

Robert Kary, president, Rice Lake Association, Iron County, WI, February 8, 2021.

Richard Thiede, secretary, Iron County Lakes and Rivers Alliance, WI, February 14, 2021.

Paul Radomski, research scientist, Minnesota Department of Natural Resources, October 3, 2020.

Adam Doll, watercraft inspection program coordinator, Minnesota Department of Natural Resources, November 19, 2020.

Paige Filice, natural resources educator, Michigan State University Extension, December 7, 2020.

Erin McFarlane, Clean Boats, Clean Waters educator, Wisconsin Department of Natural Resources, December 23, 2020.

Thomas Boisvert, aquatic invasive species coordinator, Burnett County, WI (now conservation program manager, Lincoln County, WI), November 16, 2020.

Joe Steinhage, Two Sisters Lake Property Owners Association, Oneida County, WI, October 2021.

Walt Bates, Black Oak Lake Preservation Foundation, Vilas County, WI, November 11. 2020.

Lisa Adams, Irons Area Water Protection Partners, Lake County, MI, December 16, 2021.

Dave Ferris, conservationist, Burnett County, WI, November 16, 2020.

Chris Hector, Wright County Regional Inspection Program, MN, November 18, 2020.

Sandra Swanson, McKenzie Lakes Association, Burnett County, WI, November 3, 2020.

Catherine Higley, aquatic invasive species coordinator, Vilas County, WI, November 30, 2020.

Vicki Springstead, Higgins Lake Foundation, Roscommon County, MI, February 2021.

Edgar Rudberg, CD3 Systems, St. Paul, MN, March 24, 2021.

Steve Johnson, Fish Lake Property Owners Association, Burnett County, WI, November 30, 2020.

Eric Lindberg, Environmental Sentry Protection, Maple Grove, MN, March 20, 2021.

Chapter 12

Gary Pulford, Courte Oreilles Lakes Association, Sawyer County, WI, December 22, 2020.

Brett McConnell, environmental specialist, Lac Courte Oreilles Band of Lake Superior Chippewa Indians, WI, January 11, 2021.

Michael Meyer, retired research scientist, Wisconsin Department of Natural Resources, March 16, 2021.

Alan Hamlet, associate professor of civil and environmental engineering and earth sciences, Notre Dame University, January 2021.

Carl Watras, Trout Lake Research Station, Center for Limnology, University of Wisconsin–Madison, October 2019.

Carrol Henderson, Minnesota DNR Nongame Wildlife Program, January 23, 2021.

Chapter 13

Joanne Williams, Michigan Loon Preservation Association, January 18, 2021.

Erica LeMoine, LoonWatch coordinator, Northland College, Ashland, WI, January 25, 2021.

Carrol Henderson, National Loon Center, MN, January 23, 2021.

Walter Piper, Loon Project, Oneida County, WI, January 19, 2021.

Chapter 14

Ulrik Binzer, general manager of compliance services, Granicus, March 9, 2021.

Jeffrey Goodman, planning consultant, Granicus, March 9, 2021.

Sandra Ebben, First Weber Realtors, Rhinelander, WI, March 2021.

Sandra Swanson, president, McKenzie Lakes Association, Burnett County, WI, February 20, 2021.

Michael Engleson, executive director, Wisconsin Lakes, March 23, 2021.

Chapter 15

Dave Maturen, president, Michigan Lakes and Streams Association, March 15, 2021.

Paul Radomski, research scientist, Minnesota Department of Natural Resources, April 23, 2020.

Julia Kirkwood, Michigan Department of Environment, Great Lakes, and Energy, May 15, 2020.

James Kavemeier, Tomahawk Lake Association, Oneida County, WI, March 2, 2021.

Michael Meyer, NOVA Ecological Services, WI, March 25, 2021.

Dorothy Whitmer, Gull Chain of Lakes Association, Crow Wing and Cass counties, MN, February 25, 2020.

Jeff Forester, Minnesota Lakes and River Advocates, March 15, 2021.

Bob Martini, Oneida County Lakes and Rivers Association and Wisconsin's Green Fire, March 2021.

Joe Nohner, Midwest Glacial Lakes Partnership, March 23, 2021.

Michael Engleson, executive director, Wisconsin Lakes, March 23, 2021.

Index

Aarhus University (Denmark), 166
Act 55, 57–58, 67–68, 69
Adams, Lisa, 147–48
adaptive plasticity, 118
Adirondack Center for Loon Education, 183
advocates, lake, xi, 55, 103, 202–3, 207, 218–19, 226. *See also* stewardship
Airbnb, 29, 190
AIS Early Detector Handbook, 143
Aitkin County (MN), 61
algae: blue-green, 34, 36–42, 47, 123, 167; chloride and, 200; *Daphnia* and, 113, 116, 119; filamentous, 34; golden brown, 47; green, 37, 159; phosphorus and, 76, 118; planktonic, 34–35, 42; starry stonewort, 132
Algae Adage, 39, 76
algal blooms, 43, 116, 135, 162, 166–67, 186
aluminum sulfate (alum), 42
American Sportfishing Association, 182
Anderson, Jim, 76, 81, 83, 88
Annandale (MN), 150
Antarctic ice sheets, 163
Antrim Conservation District, 46–47
Antrim County (MI), 34, 47, 62, 63, 191

aquatic invasive species. *See* invasive species, aquatic
Aquatic Invasive Species Advisory Committee, 145
Aquatic Invasive Species Awareness Week, 142
Aquatic Invasive Species Grant, 129–30
Aquatic Nuisance Species Stamps, 154
aquatic plants: critical, 212; fallen trees and, 48; invasive, 137, 142, 146–47; nutrients for, 35–37; removal of, 51, 135, 204, 213; study of, 8; wake boats and, 92, 99
Arctic sea ice, 163
artesian wells, 31
Ashland (WI), 177
Audubon Society, 186

ballast tanks, 92–93, 97, 105–6, 113, 131, 137–38, 154
Barber Lake (WI), 79–80
bass, largemouth, 52, 113, 116, 158, 160, 169–70, 171
bass, smallmouth, 51, 125, 127, 159
bass fishing, 5
Bass Lake, Town of (WI), 105
Bates, John, 13–14
Bayfield County (WI), 149

61–62; surveying for, 212; wake
boats and, 90, 92, 96–97, 98–99,
106
Eurasian water milfoil, 34, 130–31,
134–35, 145, 147–48, 153, 209, 213
European frog-bit, 132
eutrophication, 35–37, 166, 204
eutrophic lakes, 135, 204
Evangeline Township (MI), 63
evaporation, increased, 162, 165,
167–68

farms: cover crops and, 174; early,
13–14; runoff from, 33, 38–40, 83–84,
166, 224
fecal coliform bacteria, 70–73
fern-leaf pondweed, 129
Ferris, Dave, 149–50, 156
fertilizers: farm, 33, 40, 75, 166; lawn,
35, 38, 84, 185, 213, 214
Filice, Paige, 141
fires, 9–10, 13, 18
First Weber Realtors, 194
fish: chloride and, 199; condition of,
116–17; fish-finding technologies,
197; habitats, 48, 51, 52, 218–19;
overharvesting, 126, 197; sticks, 54.
See also specific types of fish
fisheries, 113, 115–17, 218
fishing gear, 120, 139
fishing line, monofilament, 183–84
fishing tackle, 179–81
Fish Lake (MN), 119
Fish Lake Property Owners Association, 156
Flambeau Lake (WI), 206
Florida, 133
flowering rush, 133
Focus on Energy (WI), 187
food webs, lake, 113, 116, 118–20, 123,
159
Forester, Jeff, 27–28, 91, 96, 107, 108,
216, 221–22

Forest Home Township (MI), 63
forestry, 14
forests, 12, 14, 172–73
frogs, green, 51

Galler, Chris, 195
glaciers, 11–12, 163
Global Climate Models (GCMs), 164
Gogebic County (MI), 58, 191
Goggin, Patrick, 46, 47–48, 50, 51, 53
Goodman, Jeff, 192, 193
Grabill, BreAnne, 34, 41, 42
Grand Traverse County (MI), 195
Granicus (company), 191–93
grants: aquatic invasive species, 141–42;
fisheries conservation, 218; hesitancy
to apply for, 213; to lakeshore property
owners, 54–55; septic system,
70; shoreline restoration, 216, 223,
226; wake boat study, 101
grass, 47, 50, 53. *See also* lawns
Greater Lake Sylvia Association
(GLSA), 150
Great Lakes, 113, 121, 127, 131, 134, 168,
205
Great Lakes AIS Landing Blitz, 142,
143
Great Lakes Commission, 142
Great Lakes Fisheries Commission,
117–18
Great Lakes Restoration Initiative, 127,
141
greenhouse gases, 162–65
Greenland ice sheets, 163
Grenzer, Linda and Kevin, 175
groundwater: about, 30–32; climate
change and, 167–68; road salt in,
199; septic systems and, 68, 71,
73–76, 78, 80
growing seasons, 64
Gull Chain of Lakes Association
(GCOLA), 102, 214–16
Gull Lake (MN), 25

Michigan Shoreland Stewards, 50
Michigan State University, 71–72, 209
Michigan State University Extension,
54, 62, 200, 204, 210
Michigan Technological University,
119–20
Michigan United Conservation clubs,
221
Michigan Waterfront Alliance, 90
Middle McKenzie Lake (WI), 111, 112
Midwest Glacial Lakes Partnership,
218
Midwest Wakesurf Association, 96
Miehls, Andrea, 118
Mille Lacs Lake (MN), 25, 117, 120,
124–26, 169
Milwaukee River, 104–5
Minnesota: home sales in, 194–95;
loon population in, 176, 178, 188;
National Lakes Assessment in, 204;
panfish in, 198; shoreland zoning in,
59–61; state aquatic invasive species
initiatives, 144–45; tourism in, 16;
walleye in, 198
Minnesota Aquatic Invasive Species
Research Center (MAISRC), 117, 123,
127–28, 137
Minnesota Arrowhead Association, 16
Minnesota Coalition of Lake Associa-
tions, 102
Minnesota Department of Revenue,
26
Minnesota Lakes and Rivers Advocates,
27, 91, 102, 107, 214, 215–16, 221
Minnesota Loon Monitoring Program,
178
Minnesota Nongame Wildlife Program,
181
Minnesota Pollution Control Agency,
84, 181, 200
Minnesota Realtors, 195
Minnocqua Chamber of Commerce
and Visitor's Bureau, 195

Minocqua Chain (WI), 25
Minocqua Lake (WI), 175
MI Paddle Stewards, 141
mitigation, shoreland, 65, 67–68
Mobile Boat Wash program, 141
monofilament fishing line, 183–84
motors, boat, 142, 149, 154
muskellunge, 111, 113, 161–62, 170
Muskie Lake (WI), 32–33
Musky Bay, 37
mystery snails, 134

National Aeronautics and Space
Administration (NASA), 162–63
National Lakes Assessment, EPA, 46,
203
National Loon Center, 181, 184
National Marine Manufacturers Asso-
ciation (NMMA), 90, 93, 98, 106,
195–96
National Oceanic and Atmospheric
Administration (NOAA), 163
Native American culture, 205
Native American tribal members, 198,
203
Native Plant Encyclopedia, 54
native plantings, 47, 50, 52–55
Natural Shoreline Partnership, Michi-
gan, 210
Natural Vegetation Waterfront Buffer
Strip provision, 63
nature preserves, 219
Nault, Michelle, 134
neighbors, lake property, 189–93
Nelson Institute for Environmental
Studies (UW–Madison), 164
Nelson Lake (WI), 79–80
New Hampshire, 106, 177, 179, 181,
188
nitrate, 74, 223
nitrogen (N), 35, 37–38, 73–74, 101,
203
Nohner, Joe, 218–19

noise pollution, 192, 193, 196
no-mow areas, 213, 214
Nonpoint Source Program, 50
North American Lake Management
  Society (NALMS), 162, 166
North Bay, 37
Northern Highland American Legion
  Forest, 20
northern pike, 113, 177
Northern Regional Lakes, 34
North Higgins Lake State Park, 152–53
North Lake (WI), 99–100, 109
North Lake Management District, 99
Northland College, 177
Northwoods Land Trust, 220
NOVA Ecological Services, 212, 217,
  218
NPK, 35, 38
nutrient pollution, 35, 135, 203
nutrients, 35–38, 43

Oakland County (MI), 58
odors, bad, 37–38, 40, 42
Ohio River, 133
oligotrophic lakes, 36, 47, 204
Oneida County (WI), 12, 16, 23, 38, 59,
  145, 176, 191, 209–10, 212, 220
Oneida County Lakes and Rivers
  Association, 41, 43, 64
Onondaga Nation, 202
Onsite Sewage Treatment Program,
  74–75
Ontario (Canada), 176, 205
Orange (CA), 182
ordinances, local: aquatic invasive
  species, 149–50; noise, 109; shore-
  land zoning and, 61–63; wake boats
  and, 108. *See also* regulations
Oregon, 106
Otsego County (MI), 195
Otter Tail, City of (MN), 70
Otter Tail County (MN), 26, 59, 70,
  83–84, 97, 154, 191, 196

Otter Tail County Land and Resource
  Management, 70
Otter Tail Lake (MN), 70, 83–86
Otter Tail Lakeshore Property Owners
  Association, 84
Otter Tail Water Management District,
  70, 83–86
overharvesting fish, 126, 197
Own Your Wake campaign, 95
oxygen, lake water: depletion, 35, 37, 41,
  52, 170–71, 173, 200; dissolved, 102;
  plants giving, 47

Pacific Ocean, 168
panfish, 169, 171, 197, 198–99
pathogens, sewage, 72–73
Pelican Lake (WI), 37–38, 39, 41–42
Pelican Lake Property Owners Association, 37, 41
per- and polyfluoroalkyl substances
  (PFAS), 221, 223
perch, yellow, 52, 116, 117, 125, 126
permits: aquatic invasive species and,
  145; building, 57, 59, 60–62, 65–66;
  herbicide, 178; septic system, 192
Personal Watercraft Industry Association (PWIA), 106
pesticides, 213, 214
Petrik, Daniel, 59–61
pet waste, 38
Phelps, Nick, 123
phosphorus (P): algal blooms and, 118,
  135, 166–67; assessment for, 203;
  eutrophication and, 35–37; in farm
  runoff, 224; as limiting nutrient,
  38–39; in MI lakes, 34–35; in MN
  lakes, 85–86; monitoring, 8; in
  property runoff, 49, 172; removing,
  42–43; sediment plumes and, 101;
  from septic systems, 70, 73, 74–76;
  sources of, 39–40; wake boats and,
  101–2; in water column, 99; in WI
  lakes, 41, 161, 204